Mechanical Properties of Polymeric, Metallic, and Composite Materials

Mechanical Properties of Polymeric, Metallic, and Composite Materials

Editors

Janusz Kluczyński
Marcin Małek

MDPI • Basel • Beijing • Wuhan • Barcelona • Belgrade • Manchester • Tokyo • Cluj • Tianjin

Editors
Janusz Kluczyński
Military University of
Technology
Warsaw, Poland

Marcin Małek
Military University of
Technology
Warsaw, Poland

Editorial Office
MDPI
St. Alban-Anlage 66
4052 Basel, Switzerland

This is a reprint of articles from the Special Issue published online in the open access journal *Materials* (ISSN 1996-1944) (available at: https://www.mdpi.com/journal/materials/special_issues/Polym_Met_Compos).

For citation purposes, cite each article independently as indicated on the article page online and as indicated below:

LastName, A.A.; LastName, B.B.; LastName, C.C. Article Title. *Journal Name* **Year**, *Volume Number*, Page Range.

ISBN 978-3-0365-8070-8 (Hbk)
ISBN 978-3-0365-8071-5 (PDF)

Cover image courtesy of Janusz Kluczyński

Contents

materials MDPI

Article

Processability of 21NiCrMo2 Steel Using the Laser Powder Bed Fusion: Selection of Process Parameters and Resulting Mechanical Properties

Jakub Łuszczek [1,*], Lucjan Śnieżek [1], Krzysztof Grzelak [1], Janusz Kluczyński [1], Janusz Torzewski [1], Ireneusz Szachogłuchowicz [1], Marcin Wachowski [1] and Marcin Karpiński [2]

1 Institute of Robots & Machine Design, Faculty of Mechanical Engineering, Military University of Technology, 00-908 Warsaw, Poland
2 Centre of Functional Materials, Lukasiewicz Research Network—Institute of Non-Ferrous Metals, 44-100 Gliwice, Poland
* Correspondence: jakub.luszczek@wat.edu.pl

Abstract: With the development and popularization of additive manufacturing, attempts have been made to implement this technology into the production processes of machine parts, including gears. In the case of the additive manufacturing of gears, the availability of dedicated materials for this type of application is low. This paper summarizes the results of research on the implementation of 21NiCrMo2 low-alloy steel, which is conventionally used to produce gears as a feedstock in the PBF-LB/M process. The work presents research on the selection of process parameters based on porosity measurements, static tensile tests, and hardness measurements. In addition, the article includes a mathematical model based on the quadratic regression model, which allows the estimation of the percentage of voids in the material depending on the assumed values of independent variables (laser power, scanning velocity, and hatch distance). The paper includes a range of process parameters that enable the production of elements made of 21NiCrMo2 steel with a density of over 99.7%. Additionally, comparative tests were carried out on PBF-LB/M-manufactured steel (in the state after printing and the state after heat treatment) and conventionally manufactured steel in terms of its mechanical and microstructural properties. The results showed that the steel exhibited similar mechanical properties to other carburizing steels (20MnCr5 and 16MnCr5) that have been used to date in PBF-LB/M processes and it can be used as an alternative to these materials.

Keywords: mechanical engineering; additive manufacturing; laser powder bed fusion; 21 NiCrMo2 steel; process parameters; post-heat treatment

Citation: Łuszczek, J.; Śnieżek, L.; Grzelak, K.; Kluczyński, J.; Torzewski, J.; Szachogłuchowicz, I.; Wachowski, M.; Karpiński, M. Processability of 21NiCrMo2 Steel Using the Laser Powder Bed Fusion: Selection of Process Parameters and Resulting Mechanical Properties. *Materials* **2022**, *15*, 8972. https://doi.org/10.3390/ma15248972

Academic Editor: Gregory N. Haidemenopoulos

Received: 18 November 2022
Accepted: 13 December 2022
Published: 15 December 2022

1. Introduction

Additive manufacturing (AM) (according to ISO/ASTM 52900:2021) is a process used for the joining of materials based on 3D CAD data. In contrast to subtractive methods, this technology is characterized by building parts layer-by-layer. The inclusion of these data in the standardized definition is the main advantage of AM technology. Hence, it is possible to obtain complex geometries, such as internal conformal cooling channels and lattice structures, which are difficult to produce with the use of conventional manufacturing technologies. In the manufacturing of machine parts via AM, powder bed fusion (PBF) (ISO/ASTM 52900) [1] technologies are most commonly used. It is worth noting that in numerous studies, the most popular method has involved the use of a laser energy source. This method is referred to as laser-based powder bed fusion of metals (PBF-LB/M)—ISO/ASTM 52911-1—and is popularly called laser powder bed fusion (L-PBF). The use of PBF technology in the production of machine parts must be economically justified. Many different research articles related to PBF-LB/M have discussed this issue [2–5]. The mentioned works indicate that the purchase costs of machines and materials represent the largest part of the financial expenditure in terms of production. On the one hand, Kamps et al. [6] pointed out that material expenses in the PBF-LB/M process constituted 32% of the overall cost. On the other hand, the American Gear Manufacturers Association (AGMA) [7] reported that the value was equal to 15%, dependent on the feedstock material type. In both cases, those costs are high and are included in the cost share of the powder in the overall process, as reported by Lindemann et al. [8]. The atomization process consumes most of the costs per 1 kg of powder. With the current balance of costs, the profitability of production using PBF-LB/M technology is achieved in unit production, in particular through the complex geometry of the parts produced. In order to expand the areas in which PBF-LB/M technology is applicable, it is necessary to develop new materials or implement those which are used in conventional processes, e.g., low-alloy steels.

One of the basic elements in the field of mechanical drive systems, which can be successfully produced via PBF-LB/M, is gears [9,10]. The loads acting on gears require the use of proper feedstock materials and postprocessing activities, which allow the proper strength properties to be obtained. Numerous studies have taken into account the production processes of power train components and the verification of their mechanical properties [11–14]. The main issue in these papers is the use of materials that are not typically dedicated to some exact solution. The most common commercially available metallic powders for PBF-LB/M are Ti6Al4V [14], AlSi10Mg [15], and the stainless steels 316 L [15], 420 [16], 17-4PH [13], and GP1 [17]. The number of steel grades that are intended for AM gears is limited [7]. One of the primary properties which allows the use of a given material in the PBF-LB/M process is its weldability (which is mostly dependent on its having a carbon amount below 0.3%). However, this feature can be modified by adding appropriate alloying elements. The most common conventionally used steels which have been implemented into PBF-LB/M processes are 16MnCr5 and 20MnCr5 [18–20], both of which are carburizing steels. According to the AGMA report [7], for this type of AM method, the following materials are in use: Pyrowear 53, Pyrowear 675, Ferrium C61, Ferrium C64, 100Cr6, M50 NIL, AISI 8620, and AISI 9310. Additionally, not all of the abovementioned materials are available in the commercial market [7]. Moreover, in the actual state of the art, some studies have been reported in which the authors used steels for quenching and tempering, i.e., 42CrMo4 [21,22], 30CrNiMo8 [23], 24CrNiMo [24], and HSLA-100 [25].

For the use of construction steels (including carburizing, quenching, and tempering) in the AM process, it is necessary to provide a properly prepared powder and the detailed development of process parameters. In the case of 16MnCr5 carburizing steel, Schmitt et al. [20,26] proved that it is possible to use AM for the production of gears without any internal cracks and imperfections and to maintain a high material density above 99.5% by means of an energy density above 100 J/mm^3 (for 99.94% it is necessary to use 108.3 J/mm^3). With such high densities, the main defect was regular shape porosity, described as gas porosity. The authors highlighted that when using the proper energy density value with different combinations of process parameters (laser power, exposure velocity, and hatching distance) they observed changes in porosity and strength properties. The ultimate tensile strength (UTS) of the obtained parts was equal to 1050 MPa and had the same value as yield strength (YS), which indicates that there was no plastic deformation and thus also no work hardening. Before the heat treatment, the conventionally manufactured material was characterized by UTS = 715 MPa and YS = 591 MPa, which represent lower values than the AM material, at 32% and 43.7%, respectively. The heat treatment (stress relief annealing) conditions for AM parts were the same as those used for the conventionally made materials. This approach resulted in a drop in the mechanical properties (UTSATH = 730 MPa, YSATH = 658 MPa, hardness from 330 HV10 to 235 HV10); additionally, it led to the creation of a fine microstructure and the elimination of the layered structure of the material. Based on the authors' work [20,26], such changes indicate a gain in the recrystallization temperature. Additionally, microscopic investigation revealed structural heterogeneities as "white areas" after nital etching. Kamps et al. [27] and Scheitler et al. [28]

proved the existence of this kind of structural anomaly in additively manufactured steel parts. Kamps et al. [27] revealed that the chemical composition of the "white area" structure is similar to that of the base material. Neither of the mentioned authors described the formation mechanism of this structure. Yang and Sisson [19] additively manufactured samples with the use of 20MnCr5 carburizing steel, employing PBF-LB/M technology. However, neither production parameters nor porosity values were given. Their research focused on the influence of heat and chemical treatments on the hardness of AM parts. The measurements were conducted using two perpendicular cross-sections of the aforementioned samples. In the as-built parts, the hardness value was equal to 287 HV5 in the XY and YZ planes. At the same time, the measured value of the parent material was 189 HV5. The differences between the mentioned material types (the AM and parent material) were mostly caused by the different forms of microstructures. The parent material exhibited a ferritic-pearlitic structure, whereas the AM parts displayed a martensitic structure. The only parameters reported for the production of 20MnCr5 steel using the PBF-LB/M technique were provided in the work of Robbato et al. [18]—106 J/mm^3 (counted on base given process parameters). However, the authors of that study did not report the porosity value obtained. Based on the works related to 16MnCr5 steel, it can be concluded that the density values were similar in both cases.

In the case of quenching and tempering steel, Zumofen et al. [23] used 30CrNiMo8 in the PBF-LB/M process. The authors used the energy density at the level of 83 J/mm^3, which allowed them to obtain a density at the level of 99.76%. The remaining voids in the material structure were characterized by a spherical shape, which is a typical indicator of gas porosity. There were no cracks in the structure of the material. The UTS and YS values of the printed material in the quenched and tempered state were higher than those in the as-built state and were comparable with those of the parent material after similar heat treatment. 42CrMo2 steel was used in the PBF-LB/M process in several works [21,22]. Damon et al. [21] achieved a porosity of 0.3% using an energy density of 85 J/mm^3. No cracks were registered in the structure of the material. As in the case of 30CrNiMo8 steel, the authors obtained higher strength parameters (UTS and YS) of the steel after quenching and tempering treatment than in the as-built state. Moreover, for this material in the as-built condition, the microstructure was also characterized by an acicular, fine-grained martensitic structure.

Based on the literature review, it can be concluded that the number of low-alloy steels used in additive manufacturing is still meager. In the case of steel for carburizing, it is necessary to use a higher energy density (above 100 J/mm^3) in order to obtain a density above 99.7% than in the case of steel for quenching and tempering (around 85 J/mm^3). In both cases, the materials in the as-built state are characterized by higher hardness and strength than the parent material. This is due to their acicular, fine-grained martensitic, or martensitic-bainitic microstructure. The change in the properties of the additive manufacturing material depends on the heat treatment performed. The particular deficiencies in the broad description of the processability of steels using PBF-LB/M are related to carburizing steels. The literature lacks a description of steels that are alternatives to the widely used 16MnCr5 or 20MnCr5. In this study, 21NiCrMo2 steel was used in order to investigate its processability with the use of PBF-LB/M, as well as the mechanical properties obtained through this process. This will provide an answer as to whether 21NiCrMo2 steel can be used as an alternative to the typical carburizing steels used to date.

2. Materials and Methods

The material taken into account in this research was the 21NiCrMo2 alloy. Such steel is traditionally used for the production of machinery equipment parts, i.e., gears, shafts, etc. Because of the limited availability of such material in the form of dedicated powder for PBF-LB/M processes, conventionally made 21NiCrMo2 steel bars were subjected to the gas atomization (GA) at the Institute of Non-Ferrous Metals (Gliwice, Poland). The LD Vacuum

Technologies GmbH atomizer (Hanau, Germany) was used for powder production. The GA process parameters are shown in Table 1.

Table 1. Process parameters used in the GA.

GA Temperature (°C)	Nozzle Diameter (mm)	Material Delivery System	Spraying Pressure (Bar)	Gas Type	Feedstock Material Type
1630	6.5	Gravitational	32	Argon	Steel bar

The powder obtained during GA was sieved using a Retsch AS200 sieve shaker (Microtrac Retsch GmbH, Haan, Germany) which allowed us to separate the proper powder fraction for the PBF-LB/M process (20–63 μm). Most powder grains (indicated by red arrows) were characterized by spherical shapes with some satellites (indicated by green arrows), which is visible in Figure 1.

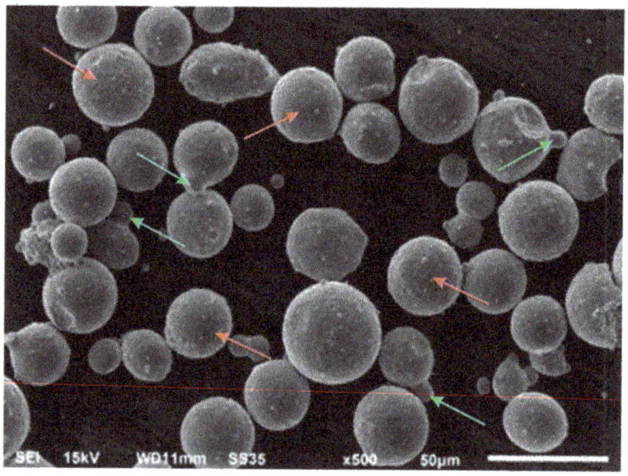

Figure 1. 21NiCrMo2 steel powder particles.

In further analyses, the material's chemical composition was investigated with the use of a scanning electron microscope (SEM) JEOL JEM-1230 (Jeol Ltd., Tokyo, Japan) equipped with an energy-dispersive spectroscopy (EDS) module. Spot measurements were made on the surfaces of the powder particles. Table 2 contains the EDS measurement results. Due to the fact that the measurements of the carbon content using the EDS method were flawed due to a large error, this value was not included in Table 2. Moreover, the authors did not have any other equipment at their disposal to conduct this type of research. During that process, there were not any material heterogeneities registered.

Table 2. The chemical composition of GA 21NiCrMo2 steel powder.

Element	Average Measurement (wt. %)	Standard Deviation (wt. %)	Parent Material (ISO) (wt. %)
C	nd	nd	0.17–0.23
Si	0.31	0.04	<0.40
Cr	0.64	0.06	0.35–0.65
Mn	0.83	0.09	0.60–0.95
Ni	0.41	0.13	0.40–0.70
Mo	0.24	0.11	0.15–0.25
Fe	97.54	0,21	Balanced

2.1. Laser Powder Bed Fusion (PBF-LB/M)

SLM 125HL (SLM Solutions GmbH, Lubeck, Germany) was used for the AM samples. The device was equipped with a 400 W single Ytterbium-fiber-laser source (wavelength 1080 nm) and a maximum scanning velocity of 10 m/s. The maximum build volume is equal to 125 × 125 × 125 mm. The range of possible layer thicknesses was 20–75 μm. The substrate plate could be heated up to 200 °C, and the AM process was performed in an argon atmosphere (in which the amount of oxygen was lower than 0.3%).

The first stage of the study involved the development of process parameters based on the porosity analysis of the AM cubic samples. Regarding the limited availability of information on the process parameters for 21NiCrMo2 steel, the default settings for H13 tool steel were used as base parameters (laser power P_L = 225 W, scanning velocity v_s = 600 mm/s, hatch distance d_H = 0.120 mm). To properly prepare for the parameter development stage, 57 different parameter groups were considered. The exact values of process parameters were tested in the following ranges: laser power P_L ranging from 160 W to 240 W (not using the full laser power due to its tendency to generate cracks in the steel structure), scanning velocity v_s ranging from 600 to 1100 mm/s, and hatching distance d_H ranging from 0.070 mm to 0.120 mm. The layer thickness l_t was kept at the same level and was equal to 0.03 mm. The platform heating was set at a value of 190 °C. As a representative parameter (dependent on P_L, v_s, d_H, and t_L) energy density E_v can be described by means of the following Equation (1):

$$E_v = \frac{P_L}{v_s \cdot d_H \cdot t_L} \left[\frac{J}{mm^3}\right] \tag{1}$$

AM samples for the porosity analysis had the form of cubes with an edge length equal to 10 mm (one sample for each parameter group). The porosity measurements were performed utilizing the Keyence VHX 7000 optical microscope (Keyence, Osaka, Japan) in both representative planes: XY($P_{\rho XY}$)—parallel to the substrate plate surface, and YZ ($P_{\rho YZ}$)—perpendicular to the substrate plate surface (Figure 2). The samples were cut using wire electrical discharge machining (WEDM) and mounted in resin for further microscopical investigation. All samples were ground using abrasive papers with a gradation from 320 to 2400 and polished using 1 μm diamond paste. As a representative porosity value, the average value was taken from $P_{\rho XY}$ and $P_{\rho YZ}$ (three different measurements for each plane).

The maximum acceptable porosity value in the entire area of measurement in cross-sections was equal to 0.3%. Additionally, to improve the development of process parameters, Design of Experiment (DOE) analysis using Statistica software 13.1 (TIBICO Software Inc., Palo Alto, Santa Clara, CA, USA) was used. For the description of mathematical porosity values, the quadratic area regression model was used. This selection was made because of the possibility of combining features of multinomial regression and fraction factorial regression models. Hence, it allowed the consideration of three independent variables and their mutual interaction. Adegok et al. [29] suggested such an approach in their work. The general form of the quadratic regression is shown in Equation (2):

$$y = \beta_0 + \beta_1 x_1 + \beta_2 x_2 + \beta_3 x_3 + \beta_{11} x_1^2 + \beta_{22} x_2^2 + \beta_{33} x_3^2 + \beta_{12} x_1 x_2 + \beta_{13} x_1 x_3 + \beta_{23} x_2 x_3 + \in \tag{2}$$

In Equation (2), y is a dependent variable (the estimated porosity value) and x_1, x_2, and x_3 are independent variables, which can be described as follows:

x_1—laser power;
x_2—scanning velocity; and
x_3—hatching distance.

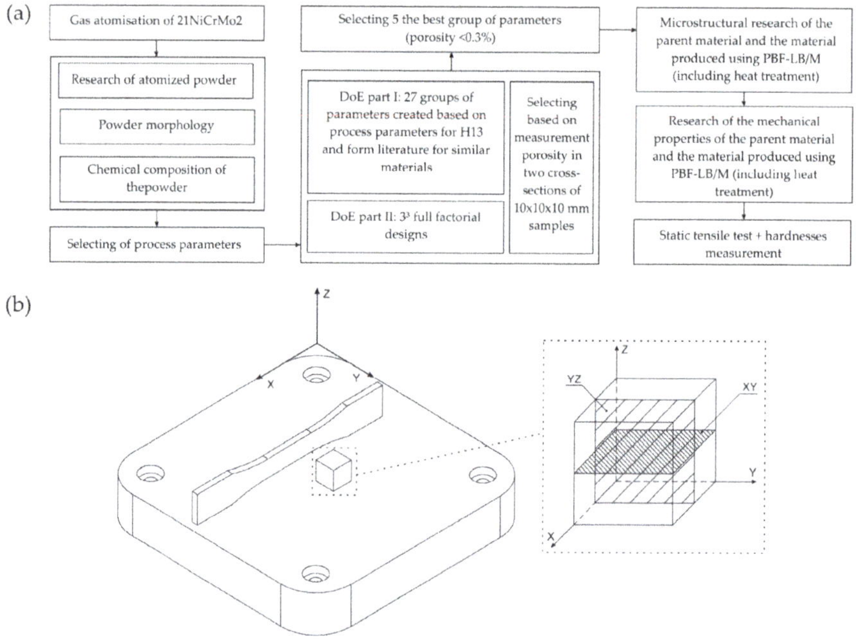

Figure 2. (**a**) Flowchart of the adopted research methodology and (**b**) the orientation of samples orientaton in the substrate plate of SLM 125HL (Z—the direction of the layers' deposition).

β_m and β_{mn} (for m = 1, 2, 3; n = 1, 2, 3) are regression coefficients, and ε is the modeling residual or error. The values of the regression coefficients were calculated with the use of the method of least squares. The calculations were divided into two parts. The first stage was dedicated to the creation of the model, based on the first porosity measurements (based on two experiments) of 27 different samples. The second stage was based on the development of further parameter combinations using the 3^3 full factorial designs (three factors: laser power, exposure speed, and hatching distance were varied at three levels for each factor). Such an approach allowed us to supplement the statistical model with the obtained results.

Additionally, the R^2 and p values were calculated. The R^2 coefficient defines how the statistical model and its predictors describe the variability of a referred parameter. The p value is the cumulative probability of drawing a sample as extreme as or more extreme than the observed one, assuming that the null hypothesis is true. The p coefficient was estimated by constructing an analysis of variance (ANOVA) table, and this was related to statistical tests. Statistical significance was set at $p < 0.05$. The validation of the obtained PBF-LB/M process parameter groups with the statistical analysis results was possible via the experimental study of the microscopic observations and porosity measurements. As a result, five process parameter groups were chosen for further research (microstructure investigation, hardness testing, and tensile tests).

2.2. Microstructure and Tensile Analysis

As-built PBF-LB/M samples and parent material parts were considered for the microstructural and tensile analyses. Because of the presence of high-temperature gradients, after the PBF-LB/M processing, all manufactured parts were subjected to stress relief annealing [30]. The lack of this operation caused deformation of samples during their separation from the substrate plate via WEDM. The heat treatment process was undertaken in a Nabertherm N11/H furnace (Nabertherm GmbH, Lilienthal, Germany). The conventionally made material was examined after normalization. All the details of the annealing process are shown in Table 3 [31]. Temperature and time values were taken from the heat treatment of conventionally made 21NiCrMo2 steel.

Table 3. 21NiCrMo2 steel heat treatment conditions [31].

Heat Treatment Type	Conditions
Stress relief annealing	Heating in the furnace until reaching 650 °C Annealing at 650 °C for 1.5 h Furnace cooling
Normalizing	Annealing in 930 °C for 4 h Air cooling

Dog-bone tensile samples were designed based on the ASTM E8 standard, and all the samples were oriented as shown in Figure 2b. This is the most favorable position for the specimens in terms of the strength of the additive manufacturing material when subjected to static tensile tests. Tensile tests were conducted on the INSTRON 8802 MTL (Instron, Norwood, MA, USA) testing machine in accordance with the PN-EN ISO 6892-1:2010 standard. Using each selected group of parameters, 5 samples were produced and tested using a tensile test. In the case of hardness testing, all measurements were made by means of a Struers DuraScan 70 system (Struers GmbH, Kopenhagen, Denmark) in accordance with the PN-EN ISO 6507-1:2007 standard. Hardness tests were conducted on the same samples, which were dedicated to the investigation of porosity. To illustrate the methodology presented here, Figure 2a shows a flowchart that briefly summarizes Section 2.

3. Results

3.1. Process Parameters and Prediction Model

The first stage of process parameter development included the AM of 27 samples. All data and predicted porosity results are included in Table A1. The modification ranges of the parameters were as follows:

- P_L from 160 to 240 W;
- v_s from 600 to 1100 mm/s; and
- h_d from 0.070 to 0.110 mm.

The groups of parameters in which the requirement $\overline{P_\rho} < 0.3\%$ was achieved were characterized by energy densities ranging from 95.2 to 121.2 J/mm^3. Additionally, we ben observed that the measured porosity values were different despite reaching similar values of energy density in the given groups. The 1.25, 1.26, and 1.27 parameter groups (Table A1) taken from the available literature [23,27] did not achieve the required porosity regime. Equation (1) was used to estimate porosity values for these cases. As variables, three parameters were considered: x_1—laser power, x_2—exposure speed, and x_3—hatching distance. Using the method of least squares, the coefficients β_m and β_{mn} (for m = 1, 2, 3; n = 1, 2, 3) were established. Additionally, by constructing the analysis of the variance table, it was possible to determine the statistical significance and create Equation (3):

$$\begin{aligned} y = 52.149000 - 0.018000x_1 - 0.067000x_2 - 525.55000x_3 + 0.00032000x_1^2 + 0.000031x_2^2 \\ +1499.389000x_3^2 - 0.000099x_1x_2 - 0.475000x_1x_3 + 0.449000x_2x_3 + 0.580000 \end{aligned} \quad (3)$$

According to a different number of considered cases for x_1, x_2, and x_3 variables, the statistical significance was valid for only one part of the equation (exposure speed, $p = 0.015$). The value of the R^2 coefficient for this exact case was equal to 0.83. Based on the established minimum of a function (3), it was possible to estimate the parameter groups, ensuring the same density as the parent material. Additionally, a determinizing estimation of the process parameters groups was made. This approach allowed us to obtain one additional group of parameters, with which the theoretical porosity value of the produced sample would be close to that produced using the extreme of function (3). The obtained process parameter groups are shown in Table 4.

Table 4. Process parameter values were obtained by following the statistical model based on the results of the first experiments.

Group	Variables	Condition
Extreme (minimum) of the function (3)	Laser power (W)	205
	Scanning velocity (mm/s)	619
	Hatch distance (mm)	0.110
Deterministically estimated group	Laser power (W)	212
	Scanning velocity (mm/s)	672
	Hatch distance (mm)	0.108

Before validating the obtained parameter groups, we decided to create 3^3 full factorial designs to distribute the measurement points equally in the considered space, which was restricted by the given variables (x_1, x_2, and x_3). Based on the obtained results (included in Table A1), it was possible to identify three process parameter value for each variable. The selection was made according to the lowest measured porosity achieved using an exact parameter group. In the case of the hatching distance, as an additional third variable, a value of 0.120 mm was taken into account.

This approach allowed for the generation of 27 different parameter groups (Table A2). Moreover, based on Equation (3), two additional groups of parameters 2.28 and 2.29 (Table 4) were estimated. Samples manufactured through the selected parameter groups were examined using the same procedure as was performed in the first part of the experiment (Section 2.1). Ten out of a total of 29 parameter groups achieved the porosity condition ($\overline{P_\rho} < 0.3\%$). Subsequently, 2.29 groups (extreme of function (4)) indicated the minimum value of porosity ($\overline{P_{\rho 2.29}} = 0.08\%$) among all the tested samples.

Another level in the selection of the process parameters in the initial statistical model was supplemented by the examination of porosity measurements. This influenced the changes in the regression and p coefficient values (Table 5). Based on the p-factor, the most significant factors in the equation among the variables were the scanning velocity and the two combinations of velocity with laser power, as well as the hatch distance.

Table 5. Regression and p coefficient values.

Coefficient Name	Coefficient Value	Value of p Coefficient
β_0	29.829000	0.152869
β_1	−0.011000	0.882832
β_{11}	0.000140	0.170741
β_2	−0.041000	0.028687
β_{22}	0.000022	0.000049
β_3	−294.833000	0.126690
β_{33}	570.583000	0.337300
β_{12}	−0.000078	0.003678
β_{13}	−0.007000	0.988815
β_{23}	0.278000	0.005153

The R^2 value equal to 0.83 did not change significantly compared to the initial value designated in the first part of this study (despite the increase in the statistical significance coefficient number). The final form of the statistical model can be described by the following Equation (4):

$$y = 29.829000 - 0.011000x_1 - 0.041000x_2 - 294.833000x_3 + 0.000140x_1^2 + 0.000022x_2^2 + 570.583000x_3^2 - 0.000078x_1x_2 - 0.007000x_1x_3 + 0.278000x_2x_3 + 0.460000 \quad (4)$$

The extreme of the function (4) has been considered in Table A2 as parameter group 2.30. However, the validation indicated a lack of repeatability in the results. This phenomenon was related to the adoption of a hatching distance that was too small, which caused overlapping of the exposure lines and, as a result, increased the number of voids. Hence, it was essential to consider all technical aspects related to the AM process manually because the statistical model did not take into account such limitations. Finally, based on the all obtained results, it was possible to identify five process parameter groups that were used in the further analysis:

- Group "2.3"—the highest value of the energy density;
- Groups "2.11" and "2.12"—the most repeatability after validation; and
- Groups "2.29" and "2.30"—the extreme function values in statistical models (3) and (4), respectively.

Statistica software was used to express the statistical model's answers as a graphical image (only when a constant value was assumed for one of the variables). The answer surfaces generated when a constant value was used for the hatching distance parameter are shown in Figure 3.

Indicating the model answer in such a way allows one to observe the range of the process parameters in which the porosity of the AM parts would be lower than 0.3%—the so-called "technological window". Such an estimation allows one to characterize the given material from the point of view of density without the need to conduct the whole spectrum of this kind of research. Additionally, five groups of process parameters selected for further research were shown on the answer surfaces (Figure 3). Process parameter groups 2.11 and 2.29 were located in the areas in which the statistical model estimated porosity values lower than 0.3%. Process parameter groups 2.3 and 2.12 were located close to the areas mentioned earlier.

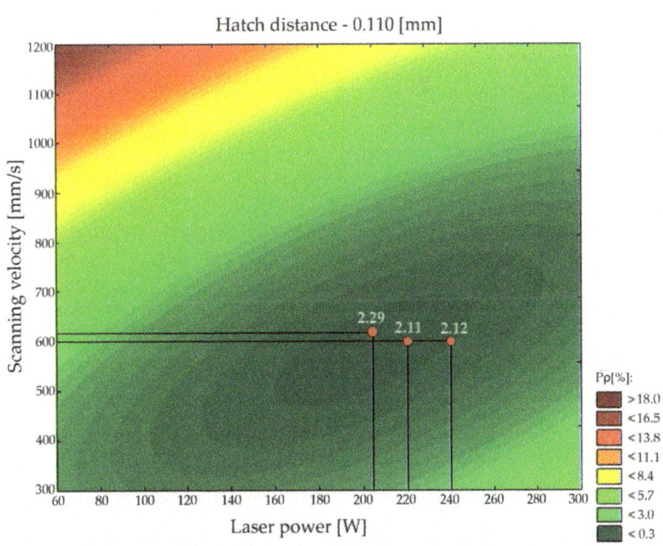

(a)

Figure 3. *Cont.*

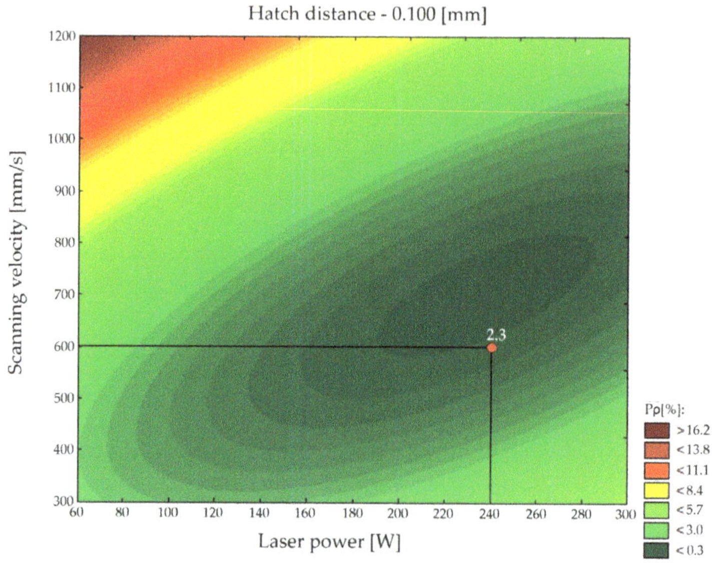

(b)

Figure 3. Answer surfaces generated for the constant value of the following hatching distance parameters: (**a**) h_d = 0.0110 mm and (**b**) h_d = 0.0100 mm.

Figure 4 shows a chart of the porosity values as a function of energy density. It can be observed that the growth of the energy density positively affected the reduction of voids in the material volume. It can also be observed that changes in the pores' shapes were related to the increase in energy density (this is shown in the figures located in the chart in Figure 4). The use of low-energy density values caused the formation of irregular void shapes, which could be connected to the "lack of fusion" phenomenon [20]. When the energy density value was greater than 80 J/mm^3, the share of spherical shapes (caused by gas porosity) and irregular shapes (caused by lack of fusion) of voids could be observed. After exceeding 100 J/mm^3, the voids mainly exhibited gas porosity characteristics. For a similar value of energy density, different authors observed similar defects in the structure of the material [20,26,27]. No cracks were found in the microstructure. The point at which the line denoting the acceptable value of the porosity crossed the trend line (determined by measurement) indicated a value of energy density equal to 104 J/mm^3. This is an empirically determined value that should be used in the AM of 21NiCrMo2 steel to achieve the condition $\overline{P_\rho}$ < 0.3%. For the experimentally indicated energy density (104 J/mm^3), the estimated porosity values with the use of Equation (4) were as follows:

- Group 2.15: (103.9 J/mm^3)—0.29%;
- Group 2.5: (104.8 J/mm^3)—0.17%.

These results led us to conclude that the statistical model created in this study enables us to estimate the porosity values of the AM parts made of 21NiCrMo2 steel.

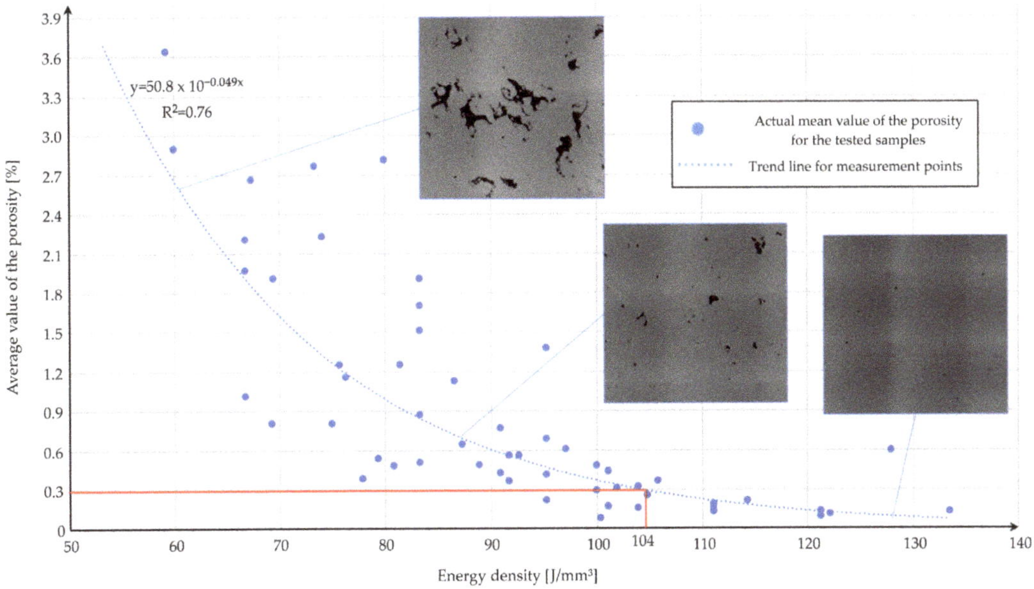

Figure 4. Porosity values for AM samples as a function of the energy density observed during the microscopical investigation.

3.2. Microstructure and Chemical Composition

PBF-LB/M and conventionally made samples were subjected to microstructural analyses (in both cross-sections). Images of the registered microstructures are shown in Figure 5. The parent material (Figure 5a) was characterized by a typical ferritic-pearlitic structure [32]. In the case of the PBF-LB/M samples, the microstructure was standard for this kind of AM process. The XY (Figure 5b) plane of the AM samples revealed exposure paths, whereas in the YZ plane, a layered structure of the deposited and melted material was visible. In addition, in Figure 5b,c, examples of porosity present in the structure are marked with green arrows. Because of the diverse cooling rates, there were observable differences in particular exposure lines, which led to a visible differentiation of the obtained microstructure. The main reason for this phenomenon is that the heat conductivity from the melt pools' outlines was more significant than that from their center, as proven by Schmitt et al. [33]. The overall microstructure showed similarity to a martensitic-bainite structure. Because of the presence of high-temperature gradients, additional heat treatment by means of stress relief was carried out (following the guidelines shown in Table 3). Hence, we registered the loss of the layered structure and made the microstructure finer but it still exhibited the form of an acicular microstructure (Figure 5d,e). These effects are typical of recrystallization annealing. The authors of [20] also demonstrated a similar phenomenon for materials produced using PBF-LB/M after the same type of heat treatment. This effect may be due to the presence of a large number of microstructural defects in the structure in its as-built state, which may lower the temperature required for the complete rebuilding of the microstructure of the material.

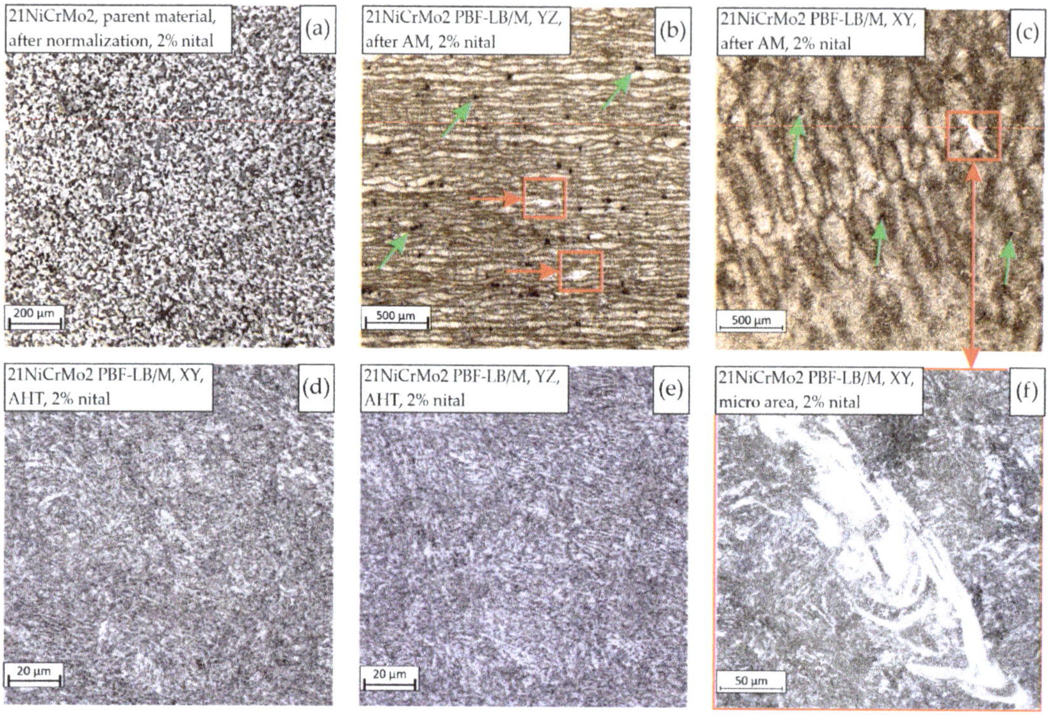

Figure 5. 21NiCrMo2 microstructure images, 2% nital etched, obtained conventionally (**a**) and via the PBF-LB/M process in both planes: XY (**b**) and YZ (**c**); after being subjected to heat treatment (AHT) and PBF-LB/M—XY (**d**), and YZ (**e**) with the visible heterogeneous micro-area of concentrated alloy elements (**f**).

Additionally, as shown in Figure 5f, in a part of the microstructure we observed a heterogeneous micro-area of concentrated alloy elements. This phenomenon was observed in both analyzed cross-sections (XY and YZ, marked with red squares and with arrows in Figure 5b,c), independently of the process parameter groups used. The dimensions of those structural heterogeneities were close to the width of a particular exposure path and a depth of 1–3 layers. According to the available research [20,27,28], such inclusions are formed during the melting process, and their chemical composition is similar to the rest of the material. In Figure 6, an EDS element distribution is identified in the mentioned micro-area.

The main alloy elements which were concentrated in the localized micro-areas were chrome, nickel, and molybdenum. The significant value of the standard deviation of the chemical composition analysis (Figure 7) indicates that a meaningful diversity of alloy elements was shared in these structures. The natural after-effect of this phenomenon was a reduction of the share of iron (Figure 6c). There is a need to conduct more detailed analyses of these phenomena in order to identify the reason for the formation of those heterogeneities.

3.3. Tensile Testing and Hardness Measurements

Dog-bone-shaped samples were made using the SLM 125HL device. The sample's orientation was selected according to Figure 2. The orientation was limited to one position only, because from the point of view of material strength, this was the most advantageous position. Some specimens were chosen to undergo stress-relief heat treatment directly after the PBF-LB/M process (without the removal of samples from the substrate plate). That kind of approach allowed us to avoid the potential deformation of the manufactured parts. Conventionally made, PBF-LB/M-produced (made using the five selected parameter groups), and heat-treated PBF-LB/M samples were tested. The tensile test results are shown in Figure 8. The porosity of samples subjected to static tensile tests was also measures and it did not differ from the values obtained in the case of cubic samples.

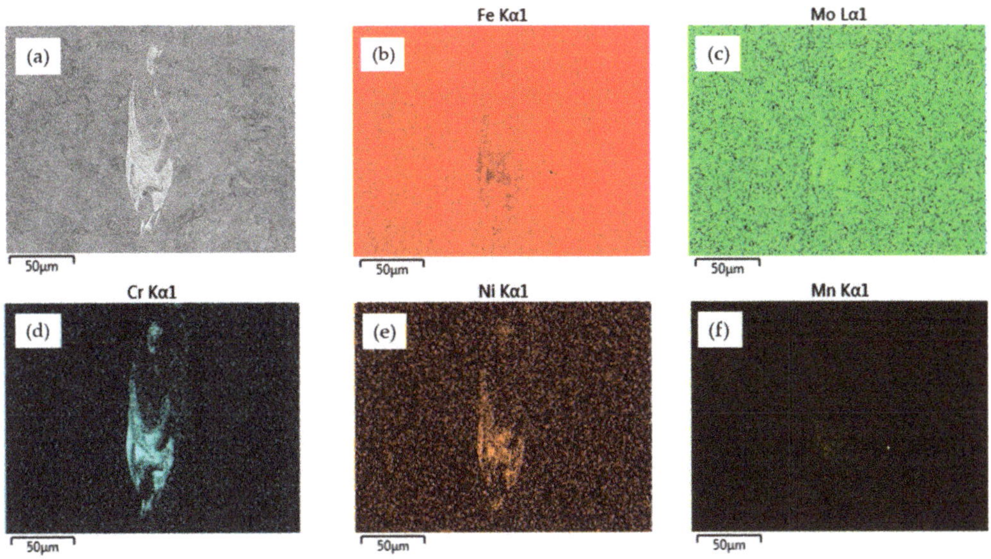

Figure 6. EDS element distribution identified in the 21NiCrMo2 steel area of structural heterogeneity: (**a**) micro-area image, (**b**) share of iron, (**c**) share of molybdenum, (**d**) share of chrome, (**e**) share of nickel, and (**f**) share of manganese.

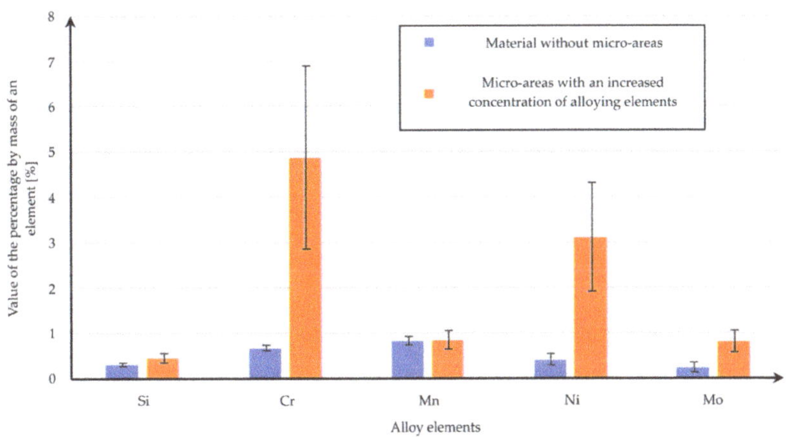

Figure 7. A bar chart of the exact alloy element percentages in the heterogeneous micro-area.

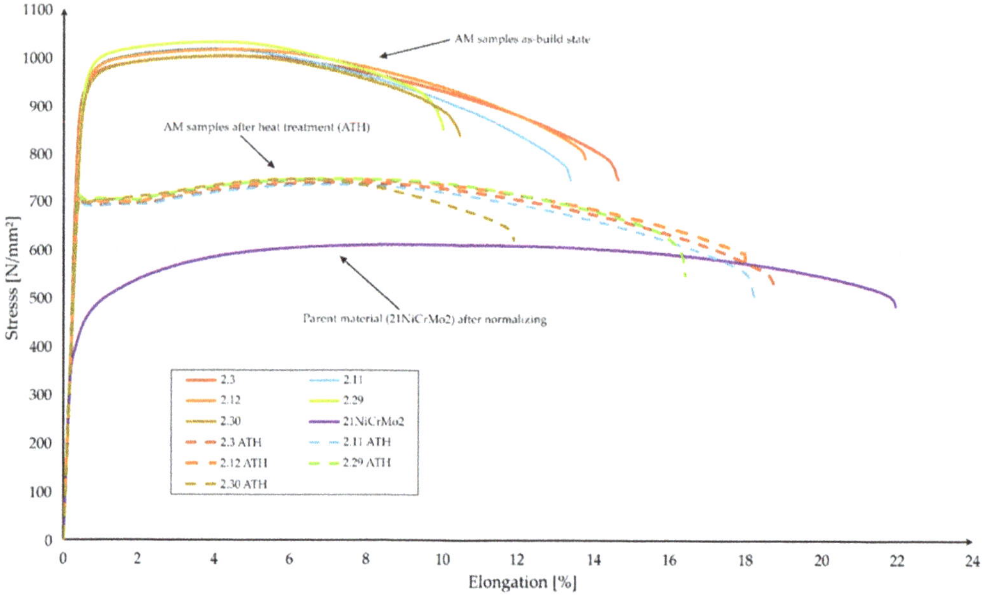

Figure 8. Stress-strain curves of the PBF-LB/M samples (as-built and after heat-treatment—AHT) compared with the parent materials (after normalization).

The PBF-LB/M-made samples, in the as-built state, were characterized by a lack of a visible yield point, independent of the process parameters used. Generally, the material in the as-built state was characterized by the lowest deformation and the highest strength, which was directly related to the martensitic-bainitic microstructure of the material. The visible difference between each parameter group was registered in the strain values during the fracturing of the samples (in the as-built state). Parts obtained with the use of parameter groups 2.30 and 2.29 were characterized by smaller strain values than those of the counterparts made with the use of groups 2.3, 2.11, and 2.12. Additional heat treatment caused a change in the tensile testing curve. Through this post-processing step, it was possible to obtain a visible yield point, but at the same time, there was a significant slope in terms of the registered UTS. Excluding the samples from group 2.30, a proportional increase in the strain of the PBF-LB/M samples was visible. Additionally, parts obtained employing the parameters from group 2.30 were characterized by the highest porosity (0.54%—Table A2), which directly affected the decrease in the total strain during tensile testing. This visibly affected the total strain values, whereas the change in UTS was negligible. The parent materials in the normalized conditions were characterized by a lower UTS and a higher strain than as-built PBF-LB/M parts. This phenomenon was strictly related to microstructural material conditions. In Table 6, we present the compiled tensile test results obtained for all tested parts.

The tensile strength of the PBF-LB/M samples was close to 1012 MPa, whereas the yield point assumed values in the range of 925–970 MPa. A slight difference between given process parameter groups was also maintained in the case of samples subjected to additional heat treatment. This indicated a minor deformation strengthening effect, which is a negative phenomenon in parts dedicated to machine design. Unlike the samples obtained using parameter group 2.30, a relatively low strain value was registered during the tensile testing of parts from group 2.29. This encourages us to constantly verify the strength properties. Additional heat treatment causes a decrease in the UTS and YS to the average values of 743 MPa and 698 MPa, respectively, and an increase in the total strain to an average of 17.3% (excluding group 2.30). Under the same conditions, the 16MnCr5 materials exhibited UTS = 730 MPa and YS = 658 MPa [26].

Table 6. Tensile test results for the parent and PBF-LB/M materials, Avr—average value, SD—standard deviation, AHT—after heat treatment.

			As-Built			
Parameter group	Avr. UTS (MPa)	SD—Avr. UTS (MPa)	Avr. YS (MPa)	SD—Avr. YS (MPa)	Avr. elong (%)	SD—Avr. elong (%)
2.3	996.1	7.8	925.0	15.0	14.1	0.6
2.12	1016.7	1.4	950.0	5.0	13.8	0.2
2.11	1023.6	4.1	965.0	5.0	13.3	1.0
2.29	1032.0	9.1	970.0	5.0	8.9	2.3
2.30	993.3	16.4	941.7	2.9	9.6	0.8
			AHT condition			
2.3 AHT	745.8	3.7	697.7	3.7	18.3	1.1
2.12 AHT	746.5	3.0	697.8	2.2	17.0	2.6
2.11 AHT	741.3	7.1	696.1	5.1	18.4	1.3
2.29 AHT	739.8	16.6	705.3	0.9	15.1	1.5
2.30 AHT	744.4	4.3	694.8	4.1	11.1	0.9
			Parent material			
21NiCrMo2	614.0	0.9	424.0	3.7	22.4	1.0

For the hardness testing, the HV1 Vickers methodology was used. A selection of the HV1 was strictly related to the minimal tip spot hole area. The registered hardness values are shown in Table 7.

Table 7. HV1 hardness measurements of the parent material and PBF-LB/M (as-built and AHT) material.

Parameter Group	HV1 (XY)	SD	HV1 (YZ)	SD
2.3	333	12	343	5
2.11	346	21	351	6
2.12	334	25	345	7
2.29	349	10	357	6
2.30	331	20	339	5
2.3 AHT	263	4	261	4
2.12 AHT	272	5	264	3
2.11 AHT	266	5	261	2
2.29 AHT	267	3	266	4
2.30 AHT	259	4	261	3
		Parent material		
21NiCrMo2	203	4	-	-

The as-built material was characterized by an average hardness of 331–357 HV1, and this was independent of the process parameter group used and the energy density used (100.3–133.3 J/mm^3). The maximal difference between measured values in both tested cross-sections (XY and YZ) was equal to 12 HV1, which could prove a lack of anisotropy in the case of this exact material property. The fact that the highest hardness of the material was observed in the as-built state was directly caused by the presence of a fine-grained martensitic-bainite structure in contrast to the ferritic-pearlitic structure of the parent material. A decrease in the hardness was registered after additional heat treatment, with the value of 260 HV1. This was the effect of microstructural changes, which also confirms the occurrence of phenomena related to recrystallization. In both as-built and AHT conditions, PBF-LB/M material was characterized by a higher hardness value than that of the parent material, which is typical in such comparisons. The material in both states was characterized by a higher hardness than PBF-LB/M 20MnCr5 and 16MnCr5 [19,20], and this may be due to the inclusion of elements which improve hardenability in 21NiCrMo2 steel.

4. Conclusions

The use of 21NiCrMo2 steel in the AM process represents a significant improvement in the number of available materials from the low-alloy steel group. The research presented here proves the possibility of using such materials in PBF-LB/M processes. The developed method of process parameter selection allowed us to identify an exact range which ensured that we could reach a very high parts density, equal to 99.7%. An essential aspect of this research is the possibility of implementing the developed mathematical model in relation to other materials dedicated to the PBF-LB/M process. Through the consideration of the obtained results, we drew the following conclusions.

1. The developed mathematical model allowed us to estimate the range of PBF-LB/M process parameters (P_L, v_s, h_d) for 21NiCrMo2 steel that can be used to obtain a material with a porosity of less than 0.3%. Moreover, this approach can be implemented for other materials when selecting the parameters of the PBF-LB/M process.
2. We empirically determined an energy density value which should be used in the PBF-LB/M process to obtain a porosity lower than 0.3%. This should be equal to 104 J/mm^3 or higher. At the same time, this value was the closest to the result obtained utilizing the mathematical model (4).
3. This value is comparable with the energy density values for other carburizing steels processed via PBF-LB/M (16MnCr5, 20MnCr5) and is higher than those used for quenching and tempering steel (30CrNiMo8, 42CrMo2).
4. A microstructural investigation allowed us to observe a share of micro-areas with alloy-element concentrations (Cr, Ni, Mo). Comparing this with the results of other studies, it can be stated that this is a phenomenon related to the low-alloy carburizing steels used in SLM technology. However, the mechanism of their formation requires deeper analysis.
5. The stress-relieving annealing temperature for the parent material produced recrystallization mechanisms in the incrementally produced material. This is evidence that for low-alloy steels, the temperatures of heat treatments should be modified in relation to those used for the parent material.
6. Samples obtained during the PBF-LB/M process with the use of selected parameter groups (energy densities between 100.3 and 133.3 J/mm^3) in as-built conditions revealed the lack of a significant yield point, an increased UTS level (on average 40% higher), and a decreased strain value (in average 24% lower) in comparison to those of the parent material.
7. Additional heat treatment in the form of stress relief caused changes in tensile testing curves (the appearance of the yield point); it also caused a reduction in the UTS from 1012 MPa to 743 MPa and an increase in total strain to 17.2% (excluding process parameter group 2.30).
8. The HV1 hardness of as-built samples assumed values in a range of 331–357 HV1, and it was higher by 140 HV1 compared to the parent material in normalized conditions (due to the presence of a martensitic-bainitic structure for 21NiCrMo2 in the as-built state). Additional heat treatment of the PBF-LB/M as-built samples caused a decrease in hardness equal to 60 HV1.
9. The obtained test results confirmed the comparable or higher strength and hardness of 21NiCrMo2 steel compared to other carburizing steels in the PBF-LB/M area. This

clearly confirms the possibility of using 21NiCrMo2 steel as an alternative to 20MnCr5 or 16MnCr5 steel in the discussed scope of research.

Further research will be focused on the influence of thermochemical treatment (carburizing hardening, and tempering) on the microstructure, as well as mechanical and fatigue properties. This is a crucial step in order to enable more advanced applied research in this field.

Author Contributions: Conceptualization, J.Ł., L.Ś. and K.G.; methodology, J.Ł., J.K., K.G. and L.Ś.; software, J.Ł. and J.K.; validation, J.Ł., J.K., J.T. and I.S.; formal analysis, J.Ł. and K.G.; investigation, J.Ł., J.T., M.W. and I.S.; resources, M.K. and I.S.; data curation, J.T., M.K. and I.S.; writing—original draft preparation, J.Ł. and J.K.; writing—review and editing, J.K. and L.Ś.; visualization, J.Ł., M.W. and I.S.; supervision, L.Ś. and K.G.; project administration, L.Ś.; funding acquisition, L.Ś. All authors have read and agreed to the published version of the manuscript.

Funding: This research was funded by the Military University of Technology, grand number: 22-757/2022.

Data Availability Statement: Not applicable.

Conflicts of Interest: The authors declare no conflict of interest. The funders had no role in the design of the study; in the collection, analyses, or interpretation of data; in the writing of the manuscript, or in the decision to publish the results.

Appendix A

Table A1. Values of process parameters for individual groups and results of measurements of the porosity of samples produced using these parameters.

Parameter Group	P_L (W)	v_s (mm/s)	h_d (mm)	t_L (mm)	E_v (J/mm^3)	$P_{\rho XY}$ (%)	$P_{\rho YZ}$ (%)	$\overline{P_\rho}$(%)
1.1	160	600	0.110	0.03	80.8	0.69	0.27	0.48
1.2	180	600	0.110	0.03	90.9	0.24	0.62	0.43
1.3	200	600	0.110	0.03	101.0	0.12	0.21	0.17
1.4	220	600	0.110	0.03	111.1	0.13	0.24	0.19
1.5	240	600	0.110	0.03	121.2	0.11	0.06	0.09
1.6	160	700	0.110	0.03	69.3	0.74	0.85	0.80
1.7	180	700	0.110	0.03	77.9	0.37	0.39	0.38
1.8	220	700	0.110	0.03	95.2	0.22	0.16	0.22
1.9	240	700	0.110	0.03	103.9	0.24	0.08	0.16
1.10	160	800	0.100	0.03	66.7	1.01	1.22	1.01
1.11	180	800	0.100	0.03	75.0	0.68	0.91	0.80
1.12	200	800	0.100	0.03	83.3	0.87	0.50	0.87
1.13	220	800	0.100	0.03	91.7	0.29	0.44	0.37
1.14	240	800	0.100	0.03	100.0	0.29	0.41	0.29
1.15	160	900	0.100	0.03	59.3	3.62	3.65	3.64
1.16	180	900	0.100	0.03	66.7	2.20	1.67	2.20
1.17	200	900	0.100	0.03	74.1	1.43	3.03	2.23
1.18	220	900	0.100	0.03	81.5	1.25	1.39	1.25
1.19	240	900	0.100	0.03	88.9	0.36	0.61	0.49
1.20	160	1000	0.100	0.03	53.3	6.07	4.52	6.07
1.21	180	1000	0.100	0.03	60.0	2.37	3.41	2.89
1.22	200	1000	0.100	0.03	66.7	1.97	2.38	1.97
1.23	220	1000	0.100	0.03	73.3	3.34	2.19	2.77
1.24	240	1000	0.100	0.03	80.0	2.81	1.71	2.81
1.25	200	900	0.070	0.03	105.8	0.39	0.34	0.37
1.26	200	900	0.110	0.03	67.3	2.52	2.79	2.66
1.27	220	1100	0.080	0.03	83.3	1.46	2.35	1.91

Appendix B

Table A2. Values of process parameters for individual groups and the measurement and prediction of results regarding the porosity of samples produced using these parameters.

Parameters Group	P_L (W)	v_s (mm/s)	h_d (mm)	t_L (mm)	E_V (J/mm^3)	$P_{\rho XY}$ (%)	$P_{\rho YZ}$ (%)	$\overline{P_\rho}$(%)—Statistic Model Equation (5)	$\overline{P_\rho}$(%)
2.1	200	600	0.100	0.03	111.1	0.13	0.16	0.26	0.15
2.2	220	600	0.100	0.03	122.2	0.12	0.09	0.26	0.11
2.3	240	600	0.100	0.03	133.3	0.13	0.12	0.28	0.13
2.4	200	700	0.100	0.03	95.2	1.30	1.45	0.32	1.38
2.5	220	700	0.100	0.03	104.8	0.27	0.23	0.17	0.25
2.6	240	700	0.100	0.03	114.3	0.30	0.12	0.18	0.21
2.7	200	800	0.100	0.03	83.3	0.63	0.38	0.83	0.51
2.8	220	800	0.100	0.03	91.7	0.64	0.47	0.53	0.56
2.9	240	800	0.100	0.03	100.0	0.62	0.34	0.53	0.48
2.10	200	600	0.110	0.03	101.0	0.53	0.34	0.17	0.44
2.11	220	600	0.110	0.03	111.1	0.14	0.11	0.17	0.13
2.12	240	600	0.110	0.03	121.2	0.14	0.12	0.38	0.13
2.13	200	700	0.110	0.03	86.6	1.15	1.08	0.50	1.12
2.14	220	700	0.110	0.03	95.2	0.78	0.60	0.35	0.69
2.15	240	700	0.110	0.03	103.9	0.28	0.30	0.29	0.32
2.16	200	800	0.110	0.03	75.8	0.58	1.92	1.28	1.25
2.17	220	800	0.110	0.03	83.3	1.28	2.11	0.98	1.70
2.18	240	800	0.110	0.03	90.9	0.71	0.82	0.78	0.77
2.19	200	600	0.120	0.03	92.6	0.59	0.52	0.18	0.56
2.20	220	600	0.120	0.03	101.9	0.29	0.32	0.18	0.31
2.21	240	600	0.120	0.03	111.1	0.21	0.12	0.29	0.17
2.22	200	700	0.120	0.03	79.4	0.49	0.57	0.79	0.53
2.23	220	700	0.120	0.03	87.3	0.56	0.72	0.64	0.64
2.24	240	700	0.120	0.03	95.2	0.42	0.39	0.59	0.41
2.25	200	800	0.120	0.03	69.4	1.80	2.02	1.86	1.91
2.26	220	800	0.120	0.03	76.4	1.07	1.23	1.54	1.15
2.27	240	800	0.120	0.03	83.3	1.08	1.93	1.35	1.51
2.28	212	672	0.108	0.03	97.1	0.99	0.22	≈0	0.61
2.29	205	619	0.110	0.03	100.3	0.10	0.06	Min. of functions (3)	0.08
2.30	240	893	0.070	0.03	127.9	0.52	0.56	Min. of functions (4)	0.54

References

1. Milewski, J.O. *Additive Manufacturing of Metals: From Fundamental Technology to Rocket Nozzles, Medical Implants, and Custom Jewelry*; Springer International Publishing: Cham, Switzerland, 2017; Volume 258. [CrossRef]
2. Rickenbacher, L.; Spierings, A.; Wegener, K. An integrated cost-model for selective laser melting (SLM). *Rapid Prototyp. J.* **2013**, *19*, 208–214. [CrossRef]
3. Taminger, K.M.; Hafley, R.A. Electron Beam Freeform Fabrication for Cost Effective Near-Net Shape Manufacturing. 2004. Available online: https://ntrs.nasa.gov/citations/20080013538 (accessed on 10 November 2020).
4. Ruffo, M.; Tuck, C.; Hague, R. Cost estimation for rapid manufacturing—Laser sintering production for low to medium volumes. *Proc. Inst. Mech. Eng. Part B J. Eng. Manuf.* **2006**, *220*, 1417–1427. [CrossRef]
5. Michael, B.; Sanjay, J.; Timothy, S.; Corey, D. Cost Modeling and Depreciation for Reused Powder Feedstocks in Powder Bed Fusion Additive Manufacturing. In Proceedings of the 27th Annual International Solid Freeform Fabrication Symposium, Austin, TX, USA, 8–10 August 2016; Volume 44, pp. 52–54.
6. Kamps, T.; Lutter-Guenther, M.; Seidel, C.; Gutowski, T.; Reinhart, G. Cost- and energy-efficient manufacture of gears by laser beam melting. *CIRP J. Manuf. Sci. Technol.* **2018**, *21*, 47–60. [CrossRef]
7. Rogers, K. *Additive Manufacturing Technologies for Gears*; American Gear Manufacturers Association: Alexandria, VA, USA, 2020.
8. Lindemann, C.; Jahnke, U.; Moi, M.; Koch, R. Analyzing product lifecycle costs for a better understanding of cost drivers in additive manufacturing. In *2012 International Solid Freeform Fabrication Symposium*; University of Texas at Austin: Austin, TX, USA, 2012; Volume 33, pp. 177–188. [CrossRef]
9. Gupta, K. Recent developments in additive manufacturing of gears: A review. *Adv. Transdiscipl. Eng.* **2018**, *8*, 131–136. [CrossRef]
10. Lin, C.; Fan, Y.; Zhang, Z.; Fu, G.; Cao, X. Additive manufacturing with secondary processing of curve-face gears. *Int. J. Adv. Manuf. Technol.* **2016**, *86*, 9–20. [CrossRef]

11. Höller, A.; Huber, F.; Zumofen, L.; Kirchheim, A.; Dinner, H.; Dennig, H.-J. Additive Manufactured and Topology Optimized Flexpin for Planetary Gears. In *Industrializing Additive Manufacturing*; Springer International Publishing: Cham, Switzerland, 2021; pp. 337–356. [CrossRef]
12. Barreiro, P.; Bronner, A.; Hoffmeister, J.; Hermes, J. New improvement opportunities through applying topology optimization combined with 3D printing to the construction of gearbox housings. *Forsch. Im Ing. Eng. Res.* **2019**, *83*, 669–681. [CrossRef]
13. Concli, F.; Bonaiti, L.; Gerosa, R.; Cortese, L.; Nalli, F.; Rosa, F.; Gorla, C. Bending fatigue behavior of 17-4 ph gears produced by additive manufacturing. *Appl. Sci.* **2021**, *11*, 3019. [CrossRef]
14. Ramadani, R.; Pal, S.; Kegl, M.; Predan, J.; Drstvenšek, I.; Pehan, S.; Belšak, A. Topology optimization and additive manufacturing in producing lightweight and low vibration gear body. *Int. J. Adv. Manuf. Technol.* **2021**, *113*, 3389–3399. [CrossRef]
15. Tezel, T.; Topal, E.S.; Kovan, V. Characterising the wear behaviour of DMLS-manufactured gears under certain operating conditions. *Wear* **2019**, *440–441*, 203106. [CrossRef]
16. Tezel, T.; Topal, E.S.; Kovan, V. Failure analysis of 3D-printed steel gears. *Eng. Fail. Anal.* **2020**, *110*, 104411. [CrossRef]
17. Rokicki, P.; Kozik, B.; Budzik, G.; Dziubek, T.; Bernaczek, J.; Przeszlowski, L.; Markowska, O.; Sobolewski, B.; Rzucidlo, A. Manufacturing of aircraft engine transmission gear with SLS (DMLS) method. *Aircr. Eng. Aerosp. Technol. Int. J.* **2015**, *3*, 397–403. [CrossRef]
18. Robatto, L.; Rego, R.; Mascheroni, J.; Kretzer, A.; Criscuolo, I.; Borille, A. Evolution of Residual Stresses induced by different L-PBF build orientations along a post-processing chain of 20MnCr5 steel. *Procedia CIRP* **2022**, *108*, 873–878. [CrossRef]
19. Yang, M.; Sisson, R.D. Carburizing Heat Treatment of Selective-Laser-Melted 20MnCr5 Steel. *J. Mater. Eng. Perform.* **2020**, *29*, 3476–3485. [CrossRef]
20. Schmitt, M.; Kamps, T.; Siglmüller, F.; Winkler, J.; Schlick, G.; Seidel, C.; Tobie, T.; Stahl, K.; Reinhart, G. Laser-based powder bed fusion of 16MnCr5 and resulting material properties. *Addit. Manuf.* **2020**, *35*, 101372. [CrossRef]
21. Damon, J.; Koch, R.; Kaiser, D.; Graf, G.; Dietrich, S.; Schulze, V. Process development and impact of intrinsic heat treatment on the mechanical performance of selective laser melted AISI 4140. *Addit. Manuf.* **2019**, *28*, 275–284. [CrossRef]
22. Jurisch, M.; Klöden, B.; Kirchner, A.; Walther, G.; Weißgärber, T. SEBM processing of 42CrMo4. *Prog. Addit. Manuf.* **2020**, *5*, 27–32. [CrossRef]
23. Zumofen, L.; Kirchheim, A.; Dennig, H.J. Laser powder bed fusion of 30CrNiMo8 steel for quenching and tempering: Examination of the processability and mechanical properties. *Prog. Addit. Manuf.* **2020**, *5*, 75–81. [CrossRef]
24. Tang, X.; Zhang, S.; Zhang, C.; Chen, J.; Zhang, J.; Liu, Y. Optimization of laser energy density and scanning strategy on the forming quality of 24CrNiMo low alloy steel manufactured by SLM. *Mater. Charact.* **2020**, *170*, 110718. [CrossRef]
25. Sridar, S.; Zhao, Y.; Li, K.; Wang, X.; Xiong, W. Post-heat treatment design for high-strength low-alloy steels processed by laser powder bed fusion. *Mater. Sci. Eng. A* **2020**, *788*, 139531. [CrossRef]
26. Schmitt, M.; Schlick, G.; Seidel, C.; Reinhart, G. Examination of the processability of 16MnCr5 by means of laser powder bed fusion. *Procedia CIRP* **2018**, *74*, 76–81. [CrossRef]
27. Kamps, T. Leichtbau von Stirnzahnrädern aus Edelstahl Mittels Laserstrahlschmelzen. Ph.D. Thesis, Technischen Universität München, Munich, Germany, 2018.
28. Scheitler, C.; Rothfelder, R.; Rasch, M.; Ahuja, B.; Schmidt, M.; Merklein, C.; Beer, O. Laser beam melting of M50NiL: Influence of inert gas flow on resulting part properties. In Proceedings of the 6th International Conference on Additive Technologies iCAT2016, Nurnberg, Germany, 29–30 November 2016.
29. Adegoke, O.; Andersson, J.; Brodin, H.; Pederson, R. Influence of laser powder bed fusion process parameters on voids, cracks, and microhardness of nickel-based superalloy alloy 247LC. *Materials* **2020**, *13*, 3770. [CrossRef] [PubMed]
30. Liu, J.; Yin, B.; Sun, Z.; Wen, P.; Zheng, Y.; Tian, Y. Hot cracking in ZK60 magnesium alloy produced by laser powder bed fusion process. *Mater. Lett.* **2021**, *301*, 130283. [CrossRef]
31. Parrish, G. *Carburizing: Microstructures and Properties*; ASM International: Almere, The Netherlands, 1999; Available online: https://books.google.pl/books?id=bAI9bDXHvZUC (accessed on 10 November 2020).
32. Boyle, E.; Northwood, D.O.; Bowers, R.; Sun, X.; Bauerle, P. The Effects of Initial Microstructure and Heat Treatment on the Core Mechanical Properties of Carburized Automotive Steels. 2018. Available online: https://www.researchgate.net/publication/286889962 (accessed on 10 November 2020).
33. Schmitt, M.; Gottwalt, A.; Winkler, J.; Tobie, T.; Schlick, G.; Stahl, K.; Tetzlaff, U.; Schilp, J.; Reinhart, G. Carbon particle in-situ alloying of the case-hardening steel 16mncr5 in laser powder bed fusion. *Metals* **2021**, *11*, 896. [CrossRef]

materials

MDPI

Article

Process Parameter Investigation and Torsional Strength Analysis of the Additively Manufactured 3D Structures Made of 20MnCr5 Steel

Bartłomiej Sarzyński *, Janusz Kluczyński, Jakub Łuszczek, Krzysztof Grzelak, Ireneusz Szachogłuchowicz, Janusz Torzewski and Lucjan Śnieżek

Institute of Robots & Machine Design, Faculty of Mechanical Engineering, Military University of Technology, Gen. S. Kaliskiego St., 00-908 Warsaw, Poland

* Correspondence: bartlomiej.sarzynski@wat.edu.pl

Abstract: An ongoing growth of the available materials dedicated to additive manufacturing (AM) significantly extends the possibilities of their usage in many applications. A very good example is 20MnCr5 steel which is very popular in conventional manufacturing technologies and shows good processability in AM processes. This research takes into account the process parameter selection and torsional strength analysis of AM cellular structures. The conducted research revealed a significant tendency for between-layer cracking which is strictly dependent on the layered structure of the material. Additionally, the highest torsional strength was registered for specimens with a honeycomb structure. To determine the best-obtained properties, in the case of the samples with cellular structures, a torque-to-mass coefficient was introduced. It indicated the best properties of honeycomb structures, which have about 10% smaller torque-to-mass coefficient values than monolithic structures (PM samples).

Keywords: additive manufacturing; torsional strength; cellular structures; 20MnCr5 Steel

Citation: Sarzyński, B.; Kluczyński, J.; Łuszczek, J.; Grzelak, K.; Szachogłuchowicz, I.; Torzewski, J.; Śnieżek, L. Process Parameter Investigation and Torsional Strength Analysis of the Additively Manufactured 3D Structures Made of 20MnCr5 Steel. *Materials* **2023**, *16*, 1877. https://doi.org/10.3390/ma16051877

Academic Editor: Andrea Di Schino

Received: 25 January 2023
Revised: 15 February 2023
Accepted: 21 February 2023
Published: 24 February 2023

1. Introduction

Additive manufacturing (AM) technologies allow the production of cellular structures, which cannot be made by other manufacturing technologies. They make it possible to achieve advantageous performance characteristics for the components and machine elements in which they are used. The main benefit of their use is their low weight, positively determining the performance generated by the device. Another important factor in favor of their use is the reduction in the amount of material used, which leads to a reduction in the cost of manufacturing the component. The ongoing growth of AM technologies requires providing significant knowledge of the behavior of the material during various types of loading and the production of cellular structures. Additionally, one of the most important advantages, based on an in situ performance property modification [1–3], is that it allows for the production of final-use parts with enhanced parameters, which are not possible to obtain with the use of conventional manufacturing technologies, i.e., molds, or drills with conformal cooling [4,5]. Furthermore, additive technology has recently shown promise for the production of components with alloys that cannot be processed with traditional processes, for example in the case of magnetic FeSi steels with a high Si content [6–8]. Many AM parts are subjected to torsional loads [9–12], which is not a very popular path in the available literature, in comparison to tensile or bending strength analysis. Such analysis is very important, especially in the case of steel powders dedicated to manufacturing high-loaded parts subjected to significant torque values. To this group of parts belong i.a. gears and shafts. Researchers have tried to implement a lattice structure to the body of these parts [13,14]. However, to generate an appropriate structure, it is important to know how the cellular structures carry the load in relation to the solid material. In the present state of the art, some works are related to the torsional strength analysis of AM parts. Halama [15] et al. showed a negative effect of additional postprocessing (machining) during torsional tests on specimens produced with the use of laser-based powder bed fusion of metal (PBF-LB/M) technology. In their research, "as built" specimens showed approximately 2/3 higher ductility than machined counterparts. The authors claimed that "as built" parts were characterized by ductile fracture when at the same time the machined specimen only had the local occurrence of ductile fractures and the prevailing features of brittle fractures. Macek et al. [16], during their analysis related to fatigue torsional tests of 18Ni300 steel, revealed a significant influence of an interlayer defect, near the surface, from which the crack nucleated into the volume of the test part. In the case of using high-pressure torsion, Han et al. [17] registered the excess number of dislocations during the nanostructuring of AM 316L stainless steel (increase by 0.27–0.32%). In another study related to torsional fatigue tests of Ti6Al4V, the authors [18] showed the shorter torsional fatigue lives of Ti6Al4V compared to the wrought material, when comparing shear strain amplitude. At the same time, AM parts were characterized by decreased cyclical softening during fatigue testing in comparison to wrought counterparts. What is more, it was found that additional annealing improves the torsional fatigue life of AM parts by more than an order of magnitude. Such a phenomenon was described by Fatemi et al. [18] and was related to the detrimental tensile residual stress relaxation generated during the AM manufacturing process, and perhaps due to increased local ductility at pores. In different research related to torsional strength analysis, Mirone et al. [19] highlighted a significant influence of surface roughness on Ti6Al4V specimens subject to torsional strength tests and registered a non-homogeneous ductile matrix mixed with brittle textures in the fracture analysis; this is a condition which reveals substantial embrittlement in fracture behavior, if compared to the static tensile loading.

In the present state of the art, there is a small amount of research related to the torsional strength analysis of AM parts obtained with the use of PBF-LB/M technology. Additionally, the availability of new steels dedicated to some exact applications (i.e., 20MnCr5 steel dedicated to carbonizing heat treatment) encourages the research of such alloys. In this research, AM lattice structures made of 20MnCr5 steel were taken into account and subjected to torsional strength tests. Such an analysis was conducted to determine material behavior during the torsional loading of thin-walled specimens.

2. Materials and Methods

Using SolidWorks 2021 software (Dassault Systems; Waltham, QC, Canada), CAD parts of test samples, including monolithic material (Figure 1), and two types of cell structures (Figures 2 and 3) were made. All specimens' geometries were based on the ISO18338:2015 standard. As a reference, a monolithic specimen was compared with all the obtained test results, the torsional strength of the material was calculated, and the most advantageous solution in terms of the strength-to-weight ratio of each component was identified.

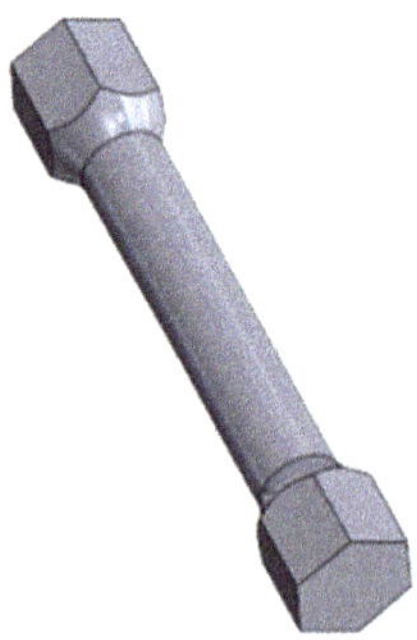

Figure 1. Monolithic specimen (PM_Samples).

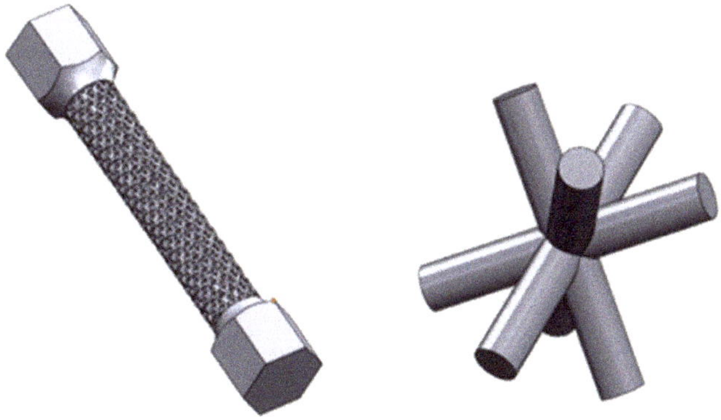

Figure 2. A view of the 'Kagome-structured' test sample (**left**) and the single cell (**right**) from which the structure was assembled (PS_Samples).

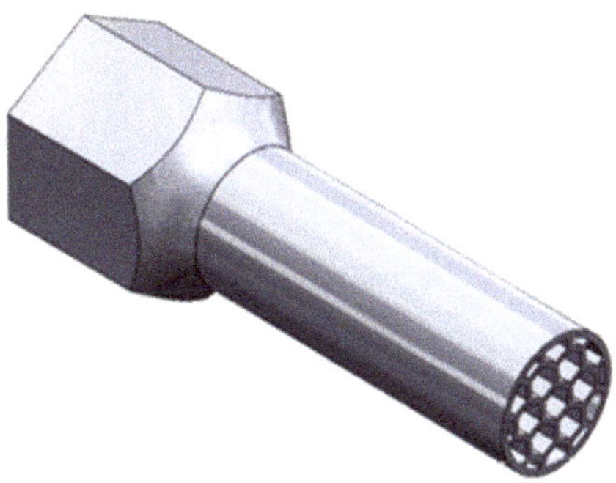

Figure 3. Half on the specimen with a honeycomb cell structure (PH Samples).

The second geometry was based on a "Kagome" structure [20] with cell diameters equal to 0.6 mm, 0.8 mm, and 1 mm. The shape of a single cell and the whole specimen is shown in Figure 2. The part was designed with a fully open structure, allowing the unmelted material to be easily removed.

A 20MnCr5 steel powder was used to make the test specimens. It is a gas-atomized, low-alloy structural steel suitable for surface hardening, which gives it a high hardness of the surface layer while allowing the core of the component to remain plastic. Table 1 shows the powder's particle diameter and the chemical composition of the steel.

Table 1. Particle size and chemical composition of 20MnCr5 and 21NiCrMo2 [21] steel powder.

Material	Powder Particles Diameter	C	Si	Cr	Mn	Ni	Mo
20MnCr5	15–45 μm	0.15%	0.19%	0.90%	1.05%	-	-
21NiCrMo2	20–63 μm	0.17–0.23%	<0.40%	0.35–0.65%	0.60–0.95%%	0.40–0.70%	0.15–0.25%

Based on our own previous research [21] on steel from the same group of alloys (21NiCrMo2), it was possible to design a parameter development matrix for 20MnCr5. Both materials are case-hardened steels with comparable chemical compositions. Therefore, it was decided to try to use similar process parameters. Moreover, low-alloy steels used in additive manufacturing are characterized by similar ranges of process parameters, regardless of the steel used [22–24]. The crucial process parameter values were the exposure velocity, laser power, and the hatching distance (clearance between exposure paths). The thickness of the powder layer was fixed with a constant value. With the component data, it was possible to calculate the energy per unit volume. For this purpose Equation (1) was used. Samples manufactured for the selection of printing parameters are shown in Figure 4. Table 2 shows the 28 groups of printing parameter values assigned to each specimen.

$$E_V = \frac{P_L}{h * v_s * H} \tag{1}$$

E_V—energy per unit volume (J/mm^3)

P_L—laser power (W)

h—hatching distance (mm)

v_s—exposure velocity (mm/s)

H—thickness of powder layer (mm)

Figure 4. Samples fabricated to determine the most beneficial printing parameters.

Table 2. Group of parameters used in the specimen manufacturing process.

Specimen Number	Laser Power (W)	Exposure Velocity (mm/s)	Hatch Spacing (mm)	Energy per Unit Volume (J/mm^3)	Thickness of Powder Layer [mm]
1	195	600	0.10	108.3	0.03
2	225	600	0.10	125.0	0.03
3	255	600	0.10	141.7	0.03
4	195	700	0.10	92.9	0.03
5	225	700	0.10	107.1	0.03
6	255	700	0.10	121.4	0.03
7	195	800	0.10	81.3	0.03
8	225	800	0.10	93.8	0.03
9	255	800	0.10	106.3	0.03
10	195	600	0.11	98.5	0.03
11	225	600	0.11	113.6	0.03
12	255	600	0.11	128.8	0.03
13	195	700	0.11	84.4	0.03
14	225	700	0.11	97.4	0.03
15	255	700	0.11	110.4	0.03
16	195	800	0.11	73.9	0.03
17	225	800	0.11	85.2	0.03
18	255	800	0.11	96.6	0.03
19	195	600	0.12	90.3	0.03
20	225	600	0.12	104.2	0.03
21	255	600	0.12	118.1	0.03
22	195	700	0.12	77.4	0.03
23	225	700	0.12	89.3	0.03

Table 2. *Cont.*

Specimen Number	Laser Power (W)	Exposure Velocity (mm/s)	Hatch Spacing (mm)	Energy per Unit Volume (J/mm^3)	Thickness of Powder Layer [mm]
24	255	700	0.12	101.2	0.03
25	195	800	0.12	67.7	0.03
26	225	800	0.12	78.1	0.03
27	255	800	0.12	88.5	0.03
28	200	1111	0.06	100.0	0.03

The Keyence (Osaka, Japan) VHX-7000 digital microscope (Figure 5) was used to analyze the quality of the sample structures that were produced using different printing parameters. Twenty-eight samples in two planes (parallel and perpendicular to layers deposition direction) were analyzed with this method. Such an approach allowed us to show the structure of the material in the melt plane (XY plane) and through the successive layers (YZ plane). The microscopic analyses for each sample were divided into two stages. In the first step, stitching of sequentially exposed areas was performed, and then the ratio of the porous area detected through the optical device to the total area of the sample was determined. The measurements for all specimens were performed using the same settings related to the detection of defects. The porosity value presented in this manuscript is based on a single measurement for each plane. The selection of the appropriate group of production parameters was made on the basis of the analysis of the results of porosity measurements. The group of parameters that allowed the production of the model element with the smallest number of pores was selected as the group used in the production of torsion test specimens.

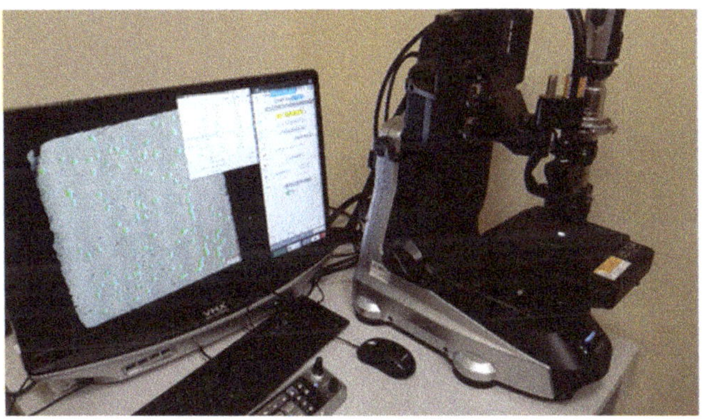

Figure 5. Keyence VHX-7000 digital microscope.

Based on the analysis of the samples' structure produced with different parameters, a group of samples with the lowest porosity was selected. During the AM process, it was planned to produce three specimens of each type:

- monolithic sample—MS,
- sample containing honeycomb structure—H,
- sample containing a bar structure with a bar diameter of 0.6 mm—S06,
- sample containing a bar structure with a bar diameter of 0.8 mm—S08,
- sample containing a bar structure with a bar diameter of 1 mm—S1.

All samples (shown in Figure 6) were produced in the vertical direction.

Figure 6. Samples after cleaning from unmelted powder.

The final stage of research was to carry out torsion tests on the manufactured specimens. For this purpose, we used the Instron 8802 testing machine presented in Figure 7 was used. The testing machine was equipped with an additional torsion-testing module designed to perform static and fatigue tests.

Figure 7. Instron 8802 torsional strength test rig.

By the means of the software integrated with the testing machine, it was possible to determine the precise character of the torsion process. The specimens were loaded at a rotational speed equal to 1°/s, (0.017 rad/s). Measuring sensors were placed in the rotating parts to register the load values of the specimen with an accuracy of 0.001 Nm. The data sampling frequency was 100 Hz.

3. Results and Discussion

3.1. AM Process Parameters Selection

The selection of the appropriate group of parameters for the production process was based on the analysis of the porosity of the samples produced with their use. Figure 8 shows an example of porosity analysis with the automatic detection of voids. This allowed the system to calculate the area identified as porous to the total measured zone. Some black dots/areas with a shade on the photo were recognized as sources of oxidation and pollution. This steel in the state as built is very susceptible to corrosion. These areas are not included in the porosity measurements.

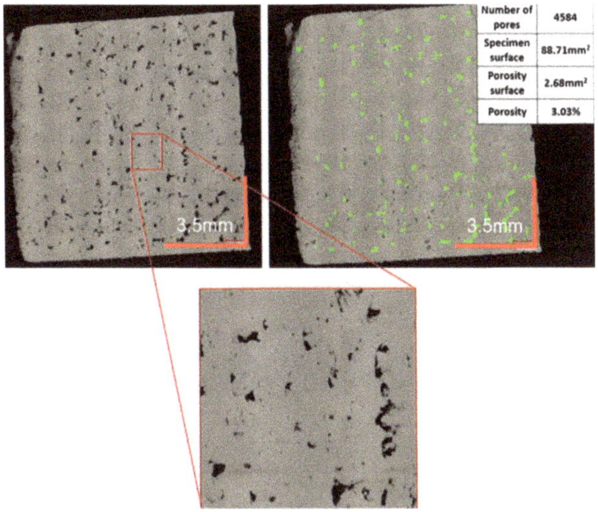

Figure 8. Digital microscope image (**left**) and the same image after porous surface analysis (green areas) on the (**right**).

Depending on the parameters used in the manufacturing process, the appearance of the specimens was significantly different. Figure 9 shows a specimen for which the twenty-seventh parameter group was used during the AM process. The laser power was 255 W, the exposure speed was 800 mm/s, and the hatching distance spacing was 0.12 mm. Based on Equation (1), the energy per unit volume was calculated and its value was 88.5 J/mm^3

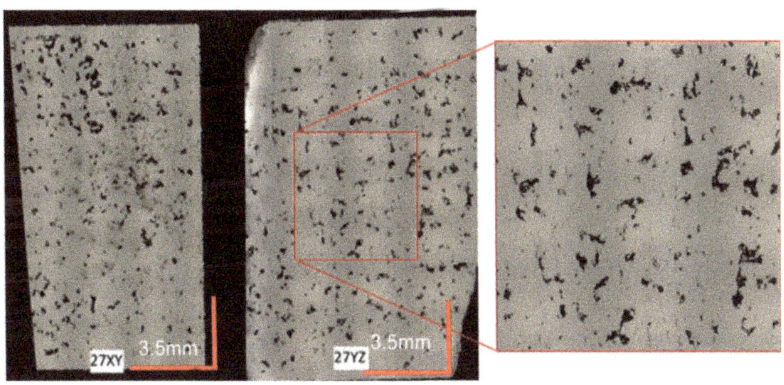

Figure 9. Image of the sample structure produced using 27th parameter group. On the left is the melt plane of a single layer. On the right is a cross-section through all layers.

Analyzing the images of the sample, a significant surface area of the porous part becomes visible immediately. The black, irregular spots indicate voids in the surface of the sample. Such defects have a very negative effect on the mechanical properties of components manufactured using a given series of parameters. During the transfer of loads through the components, stresses may be promoted in the areas concerned or notch effects may occur, leading to the formation of cracks and further material cracking. These observations were confirmed by carrying out a computer evaluation of the porous surface, which is visible in Figure 10. The surface area of defects in the sample was 3.85% in the XY plane and 4.48% in the YZ plane, giving an average of 4.17%.

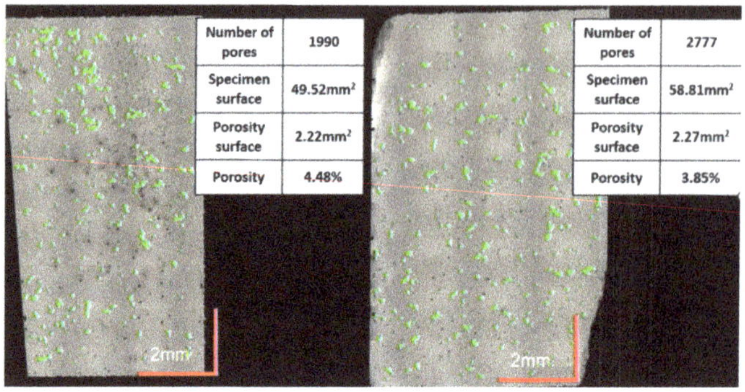

Figure 10. Presentation of porous specimen surface analysis (green color).

The minimum porous surface area of both analyzed cross-sections was registered for the 2nd parameter's group (shown in Figure 11). Process parameters in this group were as follows: a laser power equal to 225 W, an exposure velocity of 600 mm/s, and a hatching distance of 0.10 mm. The energy per unit volume was 125 J/mm^3. There are significant differences between this measurement series and the previous ones. Based on the Keyence VHX-7000 microscope software's indication of the pore areas, the number of pores was small enough to accept it (below 0.5%). This resulted in a porosity of only 0.06% in the XY plane and 0.26% in the YZ plane, giving an overall average of 0.16%. It can therefore be concluded that these parameters are appropriate for the material in question and their use will allow components with a very good quality structure to be made.

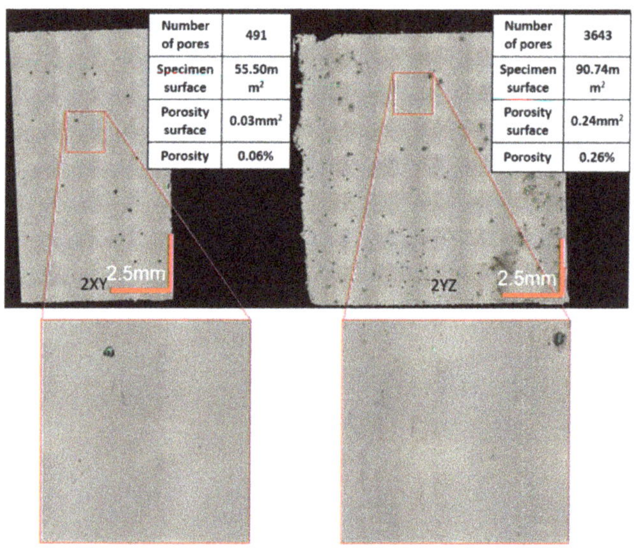

Figure 11. Evaluation of the porosity of a sample produced using 2 groups of parameters.

In order to show more clearly which samples had the highest and lowest porosity values, a bar graph is shown in Figure 12. It consists of two data series. The blue colors show the porosity values of the samples in the XY plane, while the orange colors show the porosity in the YZ plane. It is possible to read the highest porosity for samples: 7, 8, 22, and 27. The smallest porosity was registered for samples: 1, 2, 14, and 15.

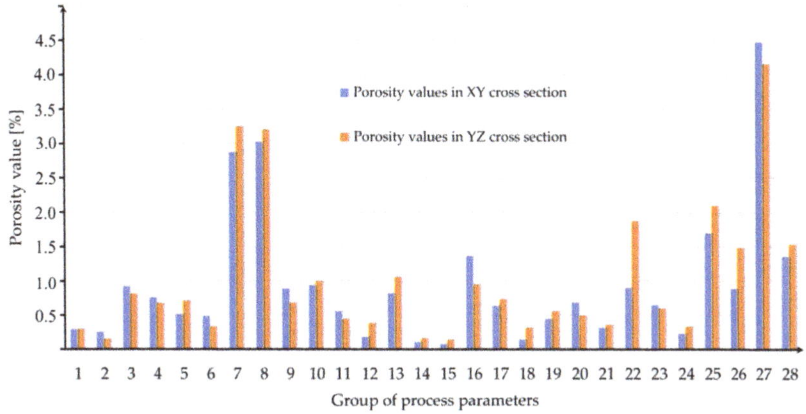

Figure 12. Columnar chart showing porosity in two planes of each tested parameter group.

Based on an analysis of the specimen structure manufactured with different parameters, a group of those that provided the lowest porosity was selected. Their values are shown in Table 3.

Table 3. Printing parameters used to manufacture cellular specimens.

Parameter and Unit	Value
Power (W)	225
Exposure velocity (mm/s)	600
Hatching distance (mm)	0.1
Layer thickness (μm)	30
Energy per unit volume (J/mm^3)	125

3.2. Torsion Tests

The torsion tests were started for monolithic specimens. Differences between the torsion angle of PM#1 and the other two cases were caused by fractured specimens PM#2 and PM#3A. The sample marked by PM#1 did not fracture during the test; the machine reached the maximum level of the torsion angle. The total rotation made by the moving part of the testing machine was 61°. At a given point, the loading torque on the specimen was 172 Nm. The stresses for all tested samples were calculated and shown in Figure 13. The similar behavior of all samples can be seen. Each line plotted through the numerical values recorded by the measuring devices is very similar. In the range of torsion angle 0–15°, the specimens deform in the elastic range, exceeding 15°, and they reach a plastic state (characteristic "flattening" of the data line). Based on the obtained results, it can be concluded that the material has a fairly low torsional strength limit.

The torsion test of the S06 specimen series indicated a similar behavior of the test samples in all three cases. Due to their very thick structure, the specimens broke at quite small angles. Specimen S06#1 failed at a torque of 9.56 Nm and a torsion angle of 4.37°. Sample S06#2 achieved the most favorable result of 9.79 Nm at an angle of 6.97°. The third specimen fractured under a torsional moment of 9.86 Nm, twisting by an angle of 6.17°. All changes in the material occurred in terms of elastic deformation. The results are shown in Figure 14. The difference between the smallest and largest value of a given parameter is only 0.3 Nm. This demonstrates the consistent failure mechanism of the specimens manufactured.

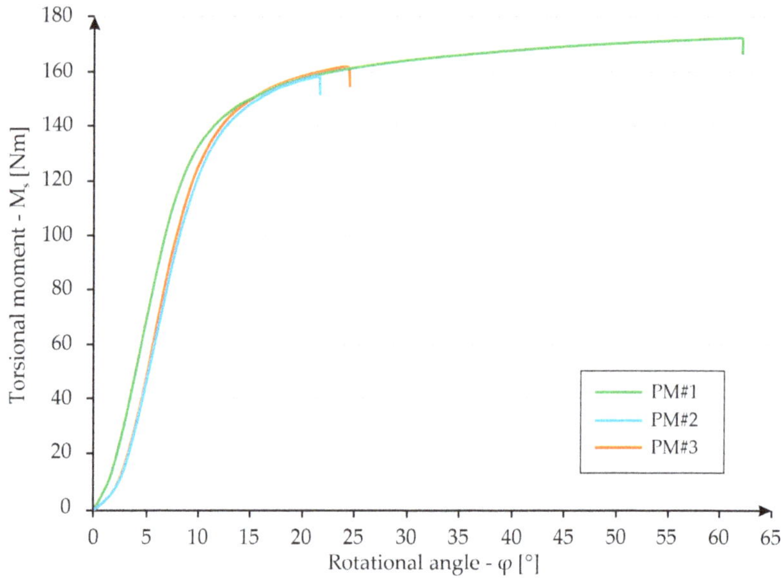

Figure 13. Graph showing the dependence of torque on torsion angle for MS specimens.

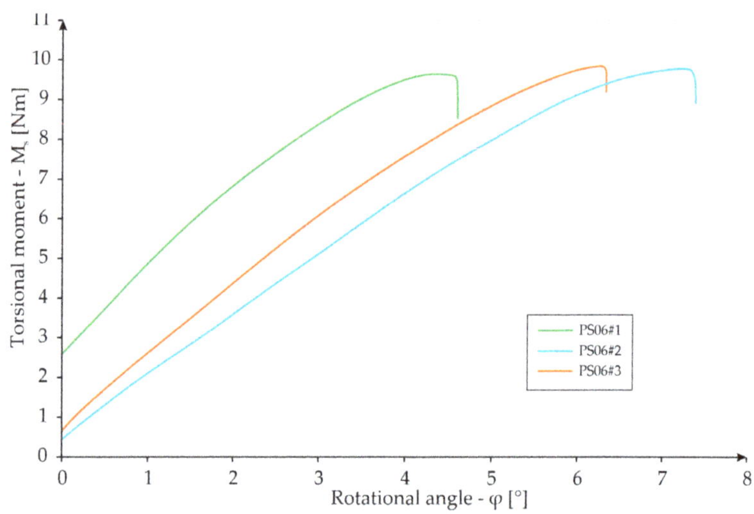

Figure 14. Graph showing the dependence of torque on torsion angle for PS06 specimens.

Sample S08#1 reached an angle of 7.3° at a torsional torque of 21 Nm. Specimen S08#2 failed at the highest value of the torsional moment. In this case, its value was 21.6 Nm, which corresponds to an angle of 7.8°. The last case retained the lowest torsion strength. Specimen S08#3 broke at an angle of 7.5° with a torsional moment of 20.3 Nm. All curves are similar to each other which indicates repeatability of all parts' properties. The angle of twist was between 7.3° and 7.8° with a torsional torque of 21 Nm and 21.6 Nm. The results are shown in Figure 15.

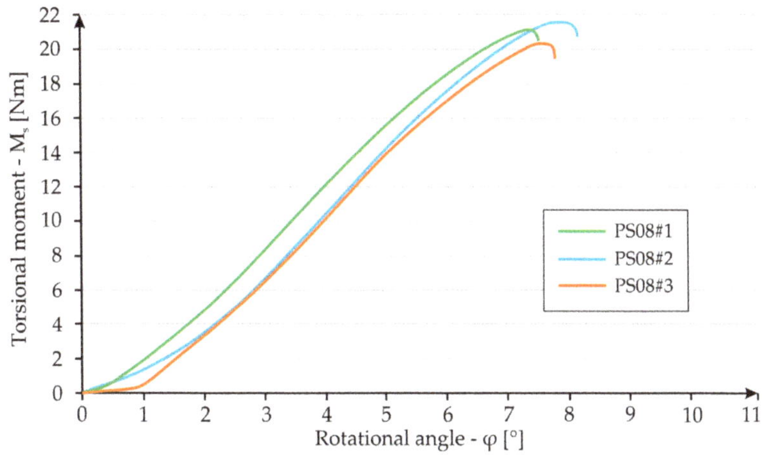

Figure 15. Graph showing the dependence of torque on torsion angle for S08 specimens.

The last of the bar structures tested was sample S1. The test results were included in the torque–torsion angle relationship shown in Figure 16. In this case, specimen S1#1 reached a torsion angle of 7.82°, breaking under a torsional torque of 38 Nm. This case achieved the highest values of the parameters described. Specimen S1#2 broke, reaching an angle of 7.1° under a torsional torque of 37 Nm. Specimen S1#3 had the lowest strength due to the torsional moment. The specimen broke, reaching 36 Nm at a torsion angle of 7.6°.

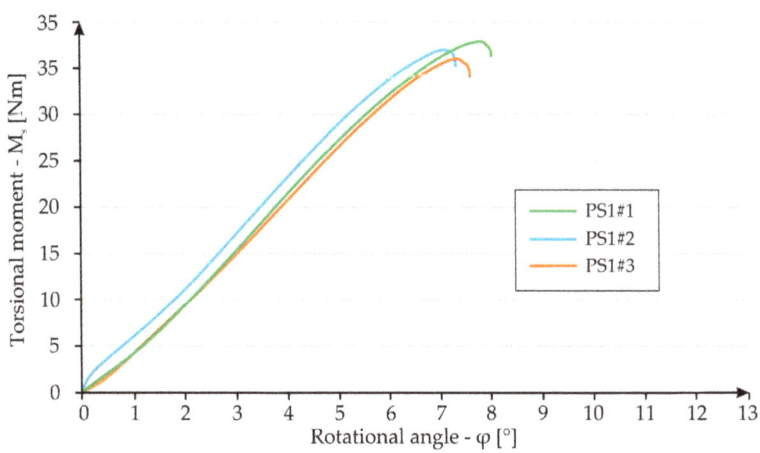

Figure 16. Graph showing the dependence of torque on torsion angle for PS1 specimens.

As a final case of cellular structure, a structure with a geometry of contacting hexagons, i.e., a "honeycomb" resemblance was tested. The test results are shown in Figure 17. In the given test series, specimen HS#1 achieved a torsion angle of 20.3° at a torsional torque of 63.8 Nm. Specimen HS#2 broke at the same torque and an angle of 18.7°. The last specimen resisted the highest load of 64.5 Nm and at an angle of 20.8°. Analyzing the graph, it is apparent that the individual data series are very similar. In the 0–8° range, all of the specimens deform elastically. When exceeding an angle value equal to 8°, the specimens indicated plastic behavior. The maximum value of the torsional moment in all three cases is strongly similar to each other.

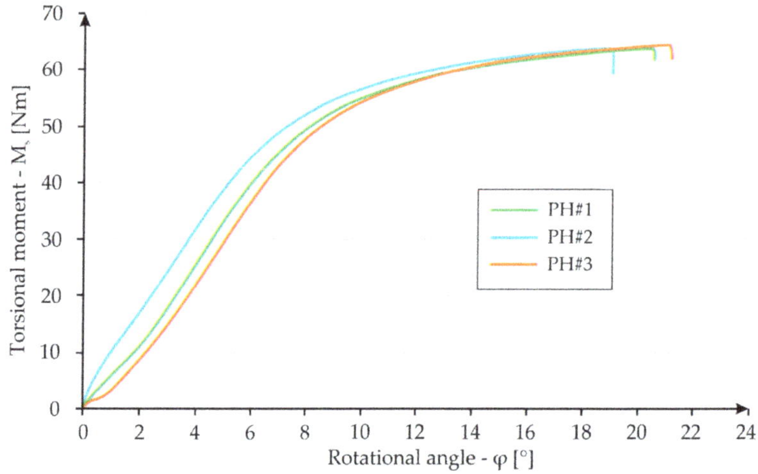

Figure 17. Graph showing the dependence of torque on torsion angle for HS specimens.

To achieve the main aim of this research project, the determination of the most favorable cell structure in terms of low mass and high strength was carried out. To this aim, a special parameter was determined using Formula (2) with the assumed torque-to-mass coefficient "W_M" being determined.

$$W_M = \frac{M_S}{m_s} \tag{2}$$

W_M—torque-to-mass coefficient,

M_S—torque at which the specimen was broken,

m_s—a mass of specimen structure.

The results of the calculations are summarized in Table 4 and the chart is shown in Figure 18.

Table 4. Registered results for different cell structures.

Specimen	Max. Torque [Nm]	Mass of Structure [g]	W_M Coefficient [Nm/kg]
PS06#1	9.65	10.67	904
PS06#2	9.79	10.42	936
PS06#3	9.86	10.64	927
PS08#1	21.09	17.94	1175
PS08#2	21.55	17.93	1202
PS08#3	20.33	17.94	1133
PS1#1	37.96	26.45	1435
PS1#2	37.01	26.48	1398
PS1#3	36.04	26.11	1380
PH#1	63.82	13.40	4764
PH#2	63.76	13.51	4720
PH#3	64.46	13.35	4828
PM#1	172.19	30.52	5641
PM#2	158.38	30.33	5222
PM#3	161.70	30.33	5332

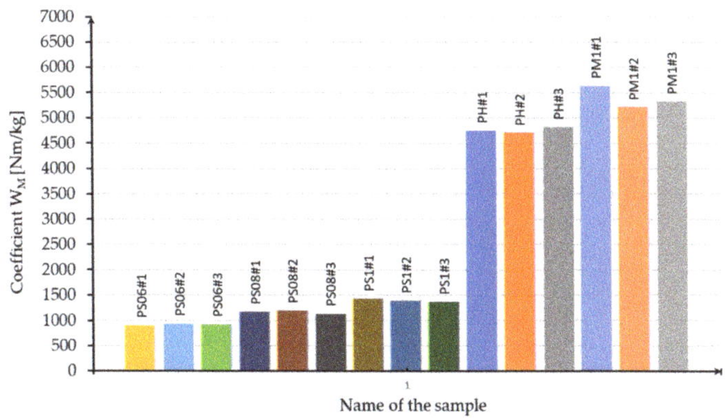

Figure 18. Columnar diagram with the values of the moment-to-torsion angle ratio for all tested samples.

Table 4 and the columnar diagram shown in Figure 18 allow us to make several observations. Of all the samples tested, the monolithic sample has the highest W_M value. The cellular structure with the highest W_M coefficient is the honeycomb structure. The highest value of the given parameter is found in sample HS#3 and is 4828 Nm/kg. Furthermore, these samples have the second smallest mass of all structure cases. Considering the samples with a "Kagome" structure, a certain relationship is apparent. As the mass of the sample increases, the W_M factor also increases. Sample S06#1 has the lowest W_M coefficient of all cases—its value is equal to 904 Nm/kg. The highest value of the described parameter among the specimens with a "Kagome" structure is found in the case of S1#1 and in the case of S1#1 is 1435 Nm/kg. Furthermore, the unit cell of the honeycomb structure has a privileged orientation related to torsional load, in contrast to the "Kagome" structure. Additionally, the area of a cross-section of the honeycomb structure is uniform throughout the whole heigh samples. The Kagome sample area of the cross-section is changeable. The PS samples always crack in the area of the cross-section that is the smallest. Figure 19 shows a chart with the dependence of torsional moment on the torsion angle for the specimens that had the highest strength in their groups. A large difference can be seen between the torsional moment values of the most favorable cellular specimens and the monolithic specimens. In this case, the cellular structures are no match for the monolithic sample. Nevertheless, the other advantages of the structures must be taken into account. When comparing the mass and the amount of material used in the manufacture, the obtained results are more favorable for samples containing cellular structures.

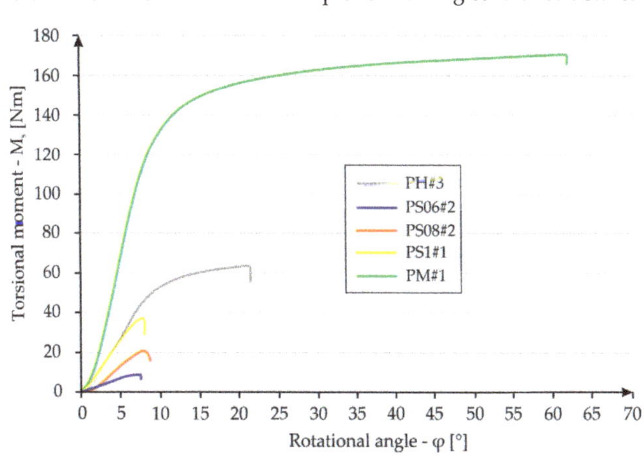

Figure 19. A chart containing the result for each specimen series.

Figure 20 presents the fractures of the tested samples. In Figure 20a, red dashed lines highlight the areas of plastic deformation of the material. The yellow color highlights the area of brittle fracture of the sample, part of which can be seen in the magnification on the right. In the case of the honeycomb structure (Figure 20b), the fracture is brittle in nature, and the specimen geometry itself has not been deformed significantly. The fractures of the specimens containing the Kagome-type structure show the occurrence of ruptures at the location of the smallest cross-sectional area, caused by the occurrence of maximum stresses at that location, regardless of the thickness of the components of the structure. The character of the breakage indicates the brittleness of the material at a given location.

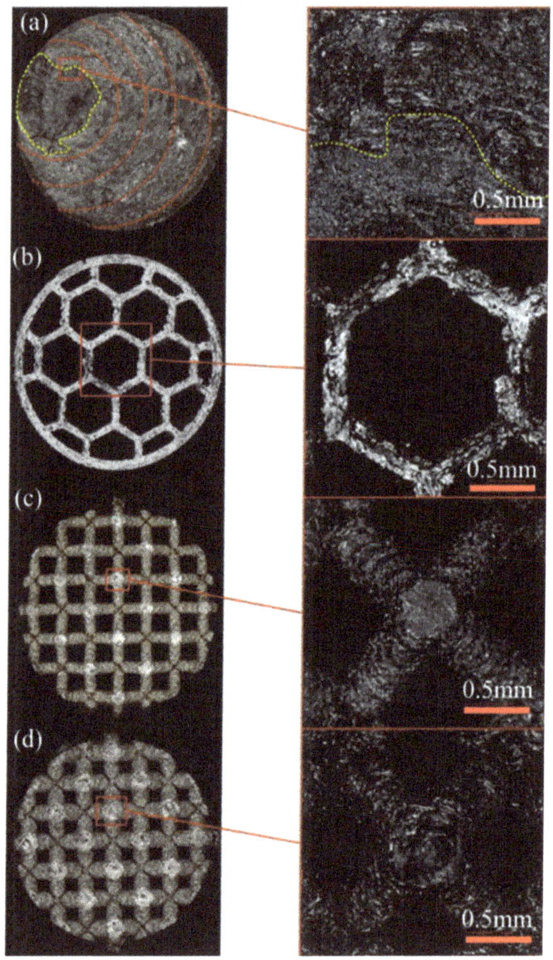

Figure 20. Images of breakthroughs: (**a**) monolithic sample, (**b**) honeycomb structure sample, (**c**) "Kagome" structure sample with 0.6 mm cell diameter and (**d**) "Kagome" structure sample with 0.8 mm cell diameter.

4. Conclusions

Based on the obtained results, the following conclusions were drawn:

1. Process parameter development for 20MnCr5 with the use of laser power equal to 225 W, an exposure velocity of 600 mm/s, and a hatching distance of 0.10 mm allowed us to obtain samples with a total porosity equal to 0.16%.

2. The total rotation angle for the monolithic sample was 61°, with the loading torque equal to 172 Nm.
3. Almost all tests (instead of PS06) indicated highly consistent results in every tested specimen.
4. To determine the best-obtained properties, in the case of the samples with cellular structures, a torque-to-mass coefficient was introduced. It indicates the best properties of honeycomb structures, which have about 10% smaller torque-to-mass coefficient values than monolithic structures (PM samples).
5. All samples broke in a plane parallel to the powder fusion layers, which is strongly affected by the part's orientation during the AM process.

Author Contributions: Conceptualization, K.G.; Methodology, K.G.; Software, J.K.; Validation, J.K., I.S. and J.T.; Resources, B.S.; Data curation, J.Ł.; Writing—original draft, B.S.; Writing—review & editing, J.K. and L.Ś. All authors have read and agreed to the published version of the manuscript.

Funding: This work was financed by the Military University of Technology, Gen. S. Kaliskiego St., 00-908 Warsaw, Poland, under research project UGB: 22-830/2022.

Institutional Review Board Statement: Not applicable.

Informed Consent Statement: Not applicable.

Data Availability Statement: Not applicable.

Conflicts of Interest: The authors declare no conflict of interest.

References

1. Pace, M.; Guarnaccio, A.; Dolce, P.; Mollica, D.; Parisi, G.; Lettino, A.; Medici, L.; Summa, V.; Ciancio, R.; Santagata, A. 3D additive manufactured 316L components microstructural features and changes induced by working life cycles. *Appl. Surf. Sci.* **2017**, *418*, 437–445. [CrossRef]
2. Labeaga-Martínez, N.; Sanjurjo-Rivo, M.; Díaz-Álvarez, J.; Martínez-Frías, J. Additive manufacturing for a Moon village. *Procedia Manuf.* **2017**, *13*, 794–801. [CrossRef]
3. Xu, K.; Li, B.; Li, S.; Luo, M.; Gao, X.; Jiang, C.; Song, L. In situ observation for the fatigue crack growth mechanism of 316L stainless steel fabricated by laser engineered net shaping. *Int. J. Fatigue* **2019**, *130*, 105272. [CrossRef]
4. Han, S.; Salvatore, F.; Rech, J.; Bajolet, J. Abrasive flow machining (AFM) finishing of conformal cooling channels created by selective laser melting (SLM). *Precis. Eng.* **2020**, *64*, 20–33. [CrossRef]
5. Kirchheim, A.; Katrodiya, Y.; Zumofen, L.; Ehrig, F.; Wick, C. Dynamic conformal cooling improves injection molding. *Int. J. Adv. Manuf. Technol.* **2021**, *114*, 107–116. [CrossRef]
6. Tiismus, H.; Kallaste, A.; Vaimann, T.; Rassõlkin, A. State of the art of additively manufactured electromagnetic materials for topology optimized electrical machines. *Addit. Manuf.* **2022**, *55*, 102778. [CrossRef]
7. Di Schino, A.; Stornelli, G. Additive Manufacturing: A New Concept for End Users. the Case of Magnetic Materials. *Acta Met. Slovaca* **2022**, *28*, 208–211. [CrossRef]
8. Stornelli, G.; Faba, A.; Di Schino, A.; Folgarait, P.; Ridolfi, M.; Cardelli, E.; Montanari, R. Properties of Additively Manufactured Electric Steel Powder Cores with Increased Si Content. *Materials* **2021**, *14*, 1489. [CrossRef]
9. Ding, Q.; Li, X.; Zhang, D.; Zhao, G.; Sun, Z. Anisotropy of poly(lactic acid)/carbon fiber composites prepared by fused deposition modeling. *J. Appl. Polym. Sci.* **2019**, *137*, 48786. [CrossRef]
10. Fatemi, A.; Molaei, R.; Phan, N. Multiaxial fatigue of additive manufactured metals: Performance, analysis, and applications. *Int. J. Fatigue* **2020**, *134*, 105479. [CrossRef]
11. Yusuf, S.M.; Nie, M.; Chen, Y.; Yang, S.; Gao, N. Microstructure and corrosion performance of 316L stainless steel fabricated by Selective Laser Melting and processed through high-pressure torsion. *J. Alloys Compd.* **2018**, *763*, 360–375. [CrossRef]
12. Ptak, M.; Mączka, M.; Gągor, A.; Sieradzki, A.; Bondzior, B.; Dereń, P.; Pawlus, S. Phase transitions and chromium(iii) luminescence in perovskite-type $[C_2H_5NH_3][Na_{0.5}Cr_xAl_{0.5-x}(HCOO)_3]$ (x = 0, 0.025, 0.5), correlated with structural, dielectric and phonon properties. *Phys. Chem. Chem. Phys.* **2016**, *18*, 29629–29640. [CrossRef] [PubMed]
13. Reinhart, G.; Teufelhart, S. Load-Adapted Design of Generative Manufactured Lattice Structures. *Phys. Procedia* **2011**, *12*, 385–392. [CrossRef]
14. Ramadani, R.; Pal, S.; Kegl, M.; Predan, J.; Drstvenšek, I.; Pehan, S.; Belšak, A. Topology optimization and additive manufacturing in producing lightweight and low vibration gear body. *Int. J. Adv. Manuf. Technol.* **2021**, *113*, 3389–3399. [CrossRef]
15. Kořínek, M.; Halama, R.; Fojtík, F.; Pagáč, M.; Krček, J.; Krzikalla, D.; Kocich, R.; Kunčická, L. Monotonic Tension-Torsion Experiments and FE Modeling on Notched Specimens Produced by SLM Technology from SS316L. *Materials* **2020**, *14*, 33. [CrossRef]

16. Macek, W.; Branco, R.; Trembacz, J.; Costa, J.; Ferreira, J.; Capela, C. Effect of multiaxial bending-torsion loading on fracture surface parameters in high-strength steels processed by conventional and additive manufacturing. *Eng. Fail. Anal.* **2020**, *118*, 104784. [CrossRef]
17. Han, J.-K.; Liu, X.; Lee, I.; Kuzminova, Y.O.; Evlashin, S.A.; Liss, K.-D.; Kawasaki, M. Structural evolution during nanostructuring of additive manufactured 316L stainless steel by high-pressure torsion. *Mater. Lett.* **2021**, *302*, 130364. [CrossRef]
18. Fatemi, A.; Molaei, R.; Sharifimehr, S.; Shamsaei, N.; Phan, N. Torsional fatigue behavior of wrought and additive manufactured Ti-6Al-4V by powder bed fusion including surface finish effect. *Int. J. Fatigue* **2017**, *99*, 187–201. [CrossRef]
19. Mirone, G.; Barbagallo, R.; Giudice, F.; Di Bella, S. Analysis and modelling of tensile and torsional behaviour at different strain rates of Ti6Al4V alloy additive manufactured by electron beam melting (EBM). *Mater. Sci. Eng. A* **2020**, *793*, 139916. [CrossRef]
20. Haydn, N.G. Wadley Multifunctional periodic cellular metals. *Philos. Trans. R. Soc. A* **2006**, *364*, 31–68.
21. Łuszczek, J.; Śnieżek, L.; Grzelak, K.; Kluczyński, J.; Torzewski, J.; Szachogłuchowicz, I.; Wachowski, M.; Karpiński, M. Processability of 21NiCrMo2 Steel Using the Laser Powder Bed Fusion: Selection of Process Parameters and Resulting Mechanical Properties. *Materials* **2022**, *15*, 8972. [CrossRef] [PubMed]
22. Damon, J.; Koch, R.; Kaiser, D.; Graf, G.; Dietrich, S.; Schulze, V. Process development and impact of intrinsic heat treatment on the mechanical performance of selective laser melted AISI 4140. *Addit. Manuf.* **2019**, *28*, 275–284. [CrossRef]
23. Jurisch, M.; Klöden, B.; Kirchner, A.; Walther, G.; Weißgärber, T. SEBM processing of 42CrMo4. *Prog. Addit. Manuf.* **2020**, *5*, 27–32. [CrossRef]
24. Zumofen, L.; Kirchheim, A.; Dennig, H.-J. Laser powder bed fusion of 30CrNiMo8 steel for quenching and tempering: Examination of the processability and mechanical properties. *Prog. Addit. Manuf.* **2020**, *5*, 75–81. [CrossRef]

materials

MDPI

Article

A Comparative Study on Laser Powder Bed Fusion of Differently Atomized 316L Stainless Steel

Krzysztof Grzelak [1], Marcin Bielecki [2], Janusz Kluczyński [1,*], Ireneusz Szachogłuchowicz [1], Lucjan Śnieżek [1], Janusz Torzewski [1], Jakub Łuszczek [1], Łukasz Słoboda [2,*], Marcin Wachowski [1], Zenon Komorek [3], Marcin Małek [4] and Justyna Zygmuntowicz [5]

1 Faculty of Mechanical Engineering, Institute of Robots & Machine Design, Military University of Technology, 2 Gen. S. Kaliskiego St., 00-908 Warsaw, Poland; krzysztof.grzelak@wat.edu.pl (K.G.); ireneusz.szachogluchowicz@wat.edu.pl (I.S.); lucjan.sniezek@wat.edu.pl (L.Ś.); janusz.torzewski@wat.edu.pl (J.T.); jakub.luszczek@wat.edu.pl (J.Ł.); marcin.wachowski@wat.edu.pl (M.W.)

2 3D LAB Sp. z o.o. Farbiarska 63B St., 02-862 Warsaw, Poland; marcin.bielecki@3d-lab.pl

3 Faculty of Materials Engineering, Institute of Materials Engineering, Military University of Technology, 2 Gen. S. Kaliskiego St., 00-908 Warsaw, Poland; zenon.komorek@wat.edu.pl

4 Faculty of Civil Engineering and Geodesy, Institute of Civil Engineering, Military University of Technology, 2 Gen. S. Kaliskiego St., 00-908 Warsaw, Poland; marcin.malek@wat.edu.pl

5 Faculty of Materials Science and Engineering, Warsaw University of Technology, 141 Woloska St., 02-507 Warsaw, Poland; justyna.zygmuntowicz@pw.edu.pl

* Correspondence: janusz.kluczynski@wat.edu.pl (J.K.); lukasz.sloboda@3d-lab.pl (Ł.S.)

Abstract: The significant growth of Additive Manufacturing (AM), visible over the last ten years, has driven an increase in demand for small gradation metallic powders of a size lower than 100 μm. Until now, most affordable powders for AM have been produced using gas atomization. Recently, a new, alternative method of powder production based on ultrasonic atomization with melting by electric arc has appeared. This paper summarizes the preliminary research results of AM samples made of two AISI 316L steel powder batches, one of which was obtained during Ultrasonic Atomization (UA) and the other during Plasma Arc Gas Atomization (PAGA). The comparison starts from powder particle statistical distribution, chemical composition analysis, density, and flowability measurements. After powder analysis, test samples were produced using AM to observe the differences in microstructure, porosity, and hardness. Finally, the test campaign covered an analysis of mechanical properties, including tensile testing with Digital Image Correlation (DIC) and Charpy's impact tests. A comparative study of parts made of ultrasonic and gas atomization powders confirms the likelihood that both methods can deliver material of similar properties.

Keywords: additive manufacturing; powder bed fusion; 316L stainless steel; ultrasonic atomization; gas atomization

Citation: Grzelak, K.; Bielecki, M.; Kluczyński, J.; Szachogłuchowicz, I.; Śnieżek, L.; Torzewski, J.; Łuszczek, J.; Słoboda, Ł.; Wachowski, M.; Komorek, Z.; et al. A Comparative Study on Laser Powder Bed Fusion of Differently Atomized 316L Stainless Steel. *Materials* **2022**, *15*, 4938. https://doi.org/10.3390/ma15144938

Academic Editor: Amir Mostafaei

Received: 23 June 2022
Accepted: 14 July 2022
Published: 15 July 2022

1. Introduction

The most popular powders on the market for additive manufacturing come from gas atomization (PAGA), which can deliver large spherical powders [1,2]. This kind of technology facilitates obtaining a high amount of proper quality powders that could be successfully used in the AM of metallic powders in Powder Bed Fusion (PBF) and (DED) technologies. Even though the growth of AM has led to the initiation of many new research paths, the need to use low amounts of metallic powder for some highly specified solutions still exists [3–5]. In the case of gas atomization devices, a lot of practical issues arise when the material type needs to be frequently changed (i.e., steel after copper alloys). One alternative and novel powder production method is based on ultrasonic atomization (UA). Such a method is commercially available on the ATO system (3D Lab Ltd., Warsaw, Poland) and utilizes capillary waves at the ultrasonic frequency to break molten metal into fine droplets [6–8].

UA is one of the most promising technologies for powder production dedicated to AM and coatings at a lower cost than PAGA, especially in the production of smaller amounts of powders (batches 1–100 kg) of customized composition (special application steels, nickel/cobalt superalloys, titanium-aluminide inter-metallics, memory shape alloys, bulk metallic glasses, high entropy alloys, noble metals, and others). UA technology facilitates obtaining spherical particles with a targeted diameter from 15 to 150 µm. Due to the selection of optimal ultrasonic frequency, the UA method can deliver a powder batch in a customized diameter range, usually ordered either in diameters of 20–60 µm, 50–120 µm, or any other narrow ranges on request. Producing powder at high efficiency in targeted diameter ranges is a crucial advantage when compared to PAGA, as deliberated later. In general, PAGA methods are characterized by the production of material particles ranging from 15 to 300 µm; classification on sieves allows for a yield of ~30% of powder in diameters needed by the AM market.

While obtaining properly shaped powder in a specified diameter range is important, it is essential to assure the proper quality of the powder. The quality control plan should address, first of all, process stability [9,10] with the lowest possible porosity [11,12], which later enables the manufacturing of mechanical parts with satisfactory mechanical properties [13–16], including fatigue properties [17–19]. The obtainment of proper particle diameters during UA processing is driven by the ejection of fine droplets of liquid metal from an ultrasonic horn (sonotrode) through ultrasonic vibrations supplied to the molten metal pool. The thermal condition in this pool is kept above liquidus temperature by continuous melting with an electric arc. The UA process is executed in argon (99.999% purity), ventilating the atomization chamber with an inlet temperature of 20 °C. To better understand the process, the main operation principles are pointed out below:

1. Raw material in the form of wire, rod, or pellets (optionally, from scrapped and milled AM print-outs "re-powder") is continuously supplied to the pressurized atomization chamber via the pushing of the material toward the sonotrode hot end, where the atomization will take place.
2. An electric arc is established between a non-consumable electrode and the sonotrode in order to melt the raw material on the sonotrode hot end to form a molten metal pool.
3. The ultrasonic vibrations are transferred through the sonotrode from its cold end, where the ultrasonic transducer is assembled, toward the hot end, and eventually to the molten metal pool. Consequently, capillary waves are formed on its surface at the same frequency as determined by the transducer. Once the magnitude of the vibrations in the pools is enough to overcome the resistance forces of the surface tension and viscosity, the capillary waves start to become unstable, and some of their crests eject the droplets at a diameter dependent on the ultrasound frequency, surface tension, and liquid metal density.
4. The droplets are ejected with some kinetic energy into the stream of the cooled inert gas. In proximity to the electric arc, the gas temperature is high enough that droplets are kept in a liquid state for a brief period, until the surface tensions round them off to almost perfect spheres.
5. Further, the droplets cool down by convection and radiation processes against the inert gas, and then solidify. The particles are conveyed with the aid of aerodynamic forces toward the atomization chamber outlet.
6. The stream of the warm inert gas with powder undergoes separation of the powder from the gas in a cyclone. Then the powder is collected below the cyclone in a sealed container.
7. The gas from the cyclone is filtered of dust (particles of a few microns), cooled, and recirculated to the process in the previous step.
8. The cooled powder is classified on various sieves to the particle sizes needed, e.g., in the range of 20–63 µm, as utilized in trials described in this paper.

In a continuous process, steps 1, 2, and 3 are run practically in the same place, i.e., the wire or rod is pushed to the sonotrode hot end. At the same time, nonstop melting keeps the molten metal pool on the sonotrode hot end, and then the ultrasonic vibrations eject the droplets from the pool—as is shown in Figure 1.

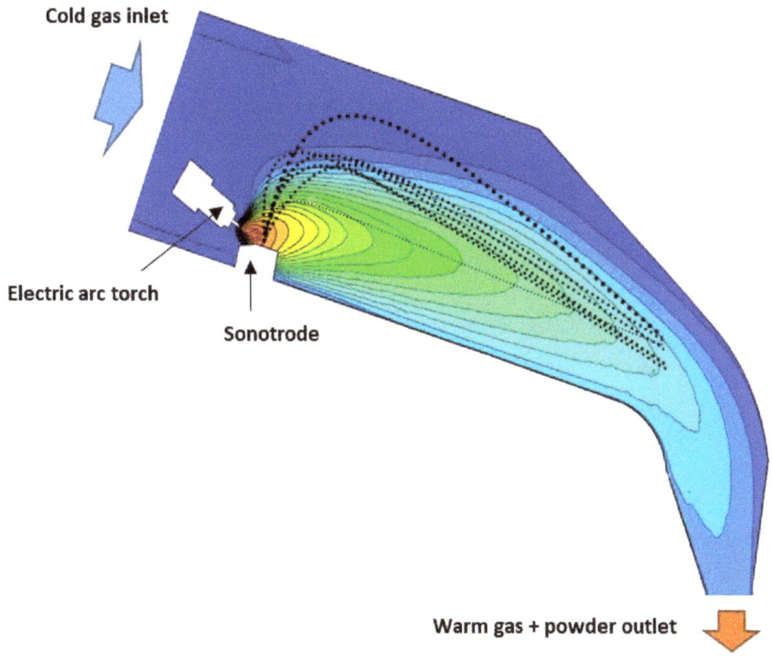

Figure 1. Atomization chamber with the temperature profile of the gas and droplet trajectories during the atomization.

Due to the ordered nature of the capillary waves driven by the ultrasonic vibrations, the output power has a narrow particle size distribution, closely dependent on the chosen ultrasound frequency. For example, for any steel, the ATO system at a frequency of 35 kHz yields powder with D50 in the range of 45–50 μm. Some minor variation depends on other factors, e.g., the heating rate (a stronger electric arc allows a higher temperature of liquid metal, hence generating finer droplets at lower viscosity and surface tensions). Due to the application of a precisely calibrated ultrasonic system and optimized electric arc heating, which allows for the creation of a relatively uniform thermal flux toward the molten metal pool, most of the powder batch has diameters close to those desired by the AM market. This physics-based phenomenon targets powder size and produces a batch with a significantly higher utilization (conversion rate) than typical gas atomization.

The main aim of this paper is to compare the UA and PAGA powders' properties, and produce sample parts with their use. Such a comparison was made to deeply analyze the differences between powder particles from each batch, and to determine how those differences affect mechanical properties of AM parts made of those two powder types. Additional discussion of all obtained results is provided.

2. Materials and Methods

2.1. Powders Utilized in a Test Campaign

The test campaign covered a comparison of two batches of AISI 316L (other specifications UNS S31703 ASTM A240 grade 316L, 1.4404) powder:

- PAGA powder (Carpenter additive, Widness, UK) produced by Plasma Arc Gas Atomization, specification CT PowderRange 316L F—this powder is one of the most popular on the AM market and widely commercially available in warehouses.

- UA powder (3D Lab Ltd., Warsaw, Poland). The powder was made of wire, commercially available for TIG/MIG/MAG welding applications—OK Autrod 317LSi (ESAB, Gothenburg, Sweden). The raw material has a low C content, which makes it particularly recommended when there is a risk of intergranular corrosion, and higher Si content, which improves wetting—advantageous in welding as well as here for the atomization process.

UA powder was produced using the ATO Plus system (3D Lab Ltd., Warsaw, Poland), which processes the raw material into powder at a frequency of 35 kHz with the electric arc set up at a current of 110 Amp. This current rating can be considered moderate if compared with a range for typical GTAW welding of stainless steels (100–160 amp for thin items).

The test campaign required 6 kg of each grade of powder, and both powders were in "fresh" conditions as delivered by their suppliers. The particle analysis and chemical composition evaluation of both powders was carried out using the Jeol JSM-6610 scanning electron microscope (SEM) Jeol JSM-6610 (Jeol, Tokyo, Japan) with an energy dispersive spectroscopy (EDS) module—Oxford X-Max.

2.2. *Powder Property Analysis*

Particle size distribution (PSD) was checked on the Keyence VHX-6000 (Keyence International, Mechelen, Belgium) digital microscope and post-processed by binary image processing (with similar quality as the laser diffraction method). Powder flowability depends on grain size distribution. The key properties of the powders for AM applications are tap and apparent densities, and flowability. These properties were measured here with the following methods:

- Apparent and tap densities—calculated according to the ASTM B212.
- Flowability—flow rate tested using a calibrated funnel on the Hall flowmeter per the ISO4490:2018 standard.

Both powders were stored in an air-filled container for a few weeks, and they were dried at 150 °C for 60 min before testing.

The chemical composition of the UA powder and its raw material was measured by an inductively coupled plasma-optical emission spectrometry PerkinElmer Optima 4300 DV (ICP-OES method) (PerkinElmer, Inc., Waltham, MA, USA). In the case of PAGA powder, the chemical composition was taken from the quality check datasheet of the purchased powder.

2.3. *Powder Melting by SLM*

Each powder batch was applied to manufacture the 3D test samples. All parts were made on the Laser—Powder Bed Fusion (L-PBF)—based, Selective Laser Melting (SLM) system—the SLM 125HL (SLM Solutions, Lubeck, Germany). To properly select the process parameters dedicated for each powder batch, 33 different laser parameters were taken into account (Table 1 (Part a–c)).

Parameters shown in the first three columns are inputs for calculating the energy density from the Formula (1).

$$\rho_E\left[\frac{J}{mm^3}\right] = \frac{L_P[W]}{e_v\left[\frac{mm}{s}\right]\cdot h_d[mm]\cdot l_t[mm]} \quad (1)$$

where:

- L_P—laser power $[W]$,
- e_v—exposure velocity [mm/s],
- h_d—hatching distance [mm],
- l_t—layer thickness [mm].

Table 1. a. Sets of analyzed process parameters for exposure velocity equal to 700 mm/s. b. Sets of analyzed process parameters for exposure velocity equal to 800 mm/s. c. Sets of analyzed process parameters for exposure velocity equal to 900 mm/s. c. Sets of analyzed process parameters for exposure velocity equal to 900 mm/s.

	a.	
No.	L_P [W]	ρ_E [J/mm^3]
1	150	59.52
2	160	63.49
3	170	67.46
4	180	71.43
5	190	75.40
6	200	79.37
7	210	83.33
8	220	87.30
9	230	91.27
10	240	95.24
11	250	99.21

	b.	
No.	L_P [W]	ρ_E [J/mm^3]
12	150	52.08
13	160	55.56
14	170	59.03
15	180	62.50
16	190	65.97
17	200	69.44
18	210	72.92
19	220	76.39
20	230	79.86
21	240	83.33
22	250	86.81

	c.	
No.	L_P [W]	ρ_E [J/mm^3]
23	150	46.30
24	160	49.38
25	170	52.47
26	180	55.56
27	190	58.64
28	200	61.73
29	210	64.81
30	220	67.90
31	230	70.99
32	240	74.07
33	250	77.16

During the research, layer thickness and hatching distance were used as the default values across all tested parameter groups, and were equal to 0.03 mm and 0.12 mm, respectively.

2.4. Structural Analysis of 3D Printed Samples

The first-level analysis was based on the sample defects by measuring their porosity and pore sizes, which are typical methods for preliminary assessment of melting performance. To select the best parameters group, cubic samples 10 × 10 × 10 mm were manufactured at set-up conditions number 1–33, as in Table 1. To allow proper porosity analysis, samples were mounted in resin, ground with 80, 320, 500, 1200, and 2000 grade abrasive papers, and polished using diamond paste (3 μm grade). Porosity was measured using the KEYENCE VHX-7000 (Keyence international, Mechelen, Belgium) digital microscope. After the process parameter selection, the samples' microstructures were evaluated. To reveal the microstructure of the samples, and acetic glycerygia solution (6 mL HCl + 4 mL HNO_3 + 4 mL CH_3COOH + 0.2 mL glycerol) was applied with an etching time of 20 s. Another structural analysis was based on Vickers hardness distribution measurements using Struers DURA SCAN 70 (Struers, Copenhagen, Denmark) hardness tester.

2.5. Tensile and Impact Testing

Samples for tensile and impact testing were manufactured with the use of parameters selected during the structural analysis. Samples were oriented horizontally (the longest dimension—along the samples' axis was oriented parallel to the substrate plate surface). Axial tensile strength tests were carried out using the Instron 8802 (Instron, Norwood, MA, USA) hydraulic pulsator. An extensometer with a measuring base of 50 mm was used for each tested sample, to allow the best quality data acquisition. During tensile testing, a digital image correlation (DIC) (non-contact, optical method) was used to measure three-dimensional (3D) deformations of the specimen surface by Dantec Dynamics (Dantec, Ulm, Germany) system. The tensile sample was compiled with the ASTM E8/E8M standard; their dimensions are shown in Figure 2.

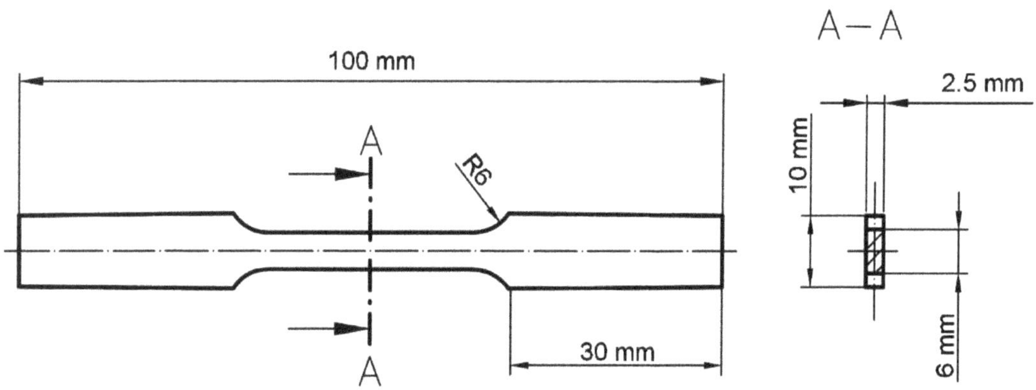

Figure 2. Dimensions of the ASTM E8\E8M standard tensile test dog-bone sample.

The impact strength tests of specimens, sized 10 × 10 × 55 mm, were carried out using the Charpy method under the EN ISO 148-1:2010 standard for KCV notch geometry. The Wolpert-Wilson PW 30 (Instron, Norwood, MA, USA) pendulum hammer was applied in the tests. The surface of the specimens was polished before the tests, and all test runs were performed at 20 °C.

3. Results and Discussion

3.1. *Powder Particle Statistical Distribution (PSD)*

PSD analysis of PAGA powder was made on the batch in as-delivered condition, i.e., grain size 15–45 µm in the range D10–D90, per production specification. The UA batch output right after production (before classification) was:

- 98.0% of raw material (wire) was converted in powder < 100 µm (i.e., particles useful for AM, melting, coatings), hence, the scrap rate is much lower than in the PAGA process because a conversion rate < 100 µm typically yields in the range of 85–90% [1,3]
- 92.3% of the powder mass can be classified in the range of 20–63 µm, i.e., applicable for SLM, which is also a much better value than for the PAGA process (typically 70–75% for steels)

Eventually, the UA powder was classified as 20–63 µm according to DIN 66165-1 with a sieve shaker brand Multiserw model LPzE-3e at a frequency of 50 Hz. The powder samples for PSD assessment contain 18400 particles for PAGA powder and 8150 for UA powder—both samples had a similar mass, as it follows the fact that PAGA powder was finer than UA. The auxiliary quality metric of the powder batch is also reported: their Span = (D90–D10)/D50 (the lower, the better), and sphericity is defined as the mean value of the proportion of Dim/Dmax for all particles in the batch as both correlated with powder flowability during placement in the melting bed. Figure 3 presents a chart that compares the cumulative distributions of the powder for PAGA and UA methods (at the standard set-up of the ATO system to maximize output). A significant disadvantage of gas atomization is its widespread diameter range [1,3,20]. However, a median particle has a size of 38–45 µm [20], but the upper value at D90 reaches 85–110 µm, with less than 70–75% of the batch (per volume) being a size of <63 µm, which is desirable for AM. The ultrasonic atomization at a frequency of 35 kHz can yield a powder with a median of 44.0 µm; hence 92% of the batch meets PBF requirements at a minimal scrap rate of the rejected powder. Although the UA output (mass rate) is less than that of the PAGA, the ATO system is capable of producing the required powder in a single run with a high conversion rate (>95% of raw material converted in unclassified powder).

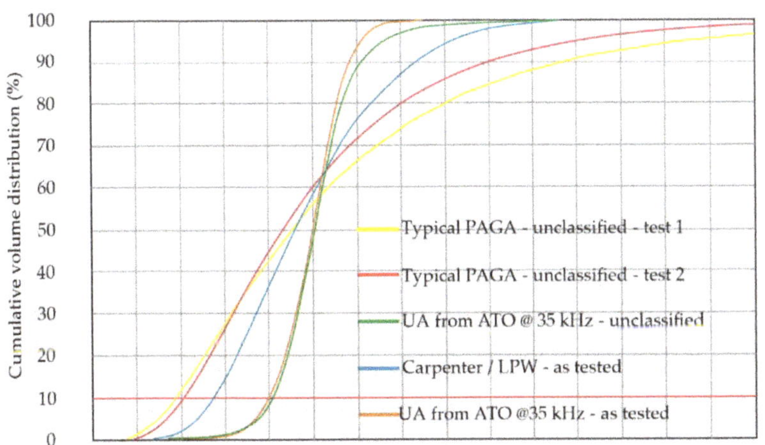

Figure 3. Cumulative statistical distribution (volume %) for the powder batches as in the test campaign, test 1 and test 2 courses were taken from the references [1,3] respectively.

In general, the UA powder shows more favorable quality metrics, i.e., lower span, more narrow statistical distribution, and higher sphericity, compared with commercially available PAGA powder. A more limited distribution (i.e., at a low span) and better sphericity of UA powder are apparent in Figure 4. The key statistical parameters are documented in Table 2 (Part A), where the PSD is based on a particle (item) counting, and Table 2 (Part B), where the PSD is evaluated based on the particle set weight. The diameters D10, D50, and D90 are measured as the longest Feret diameters. Comparing the PSD based on optical methods (like binary image processing or laser diffraction), the UA powder seems to be more coarse (median D50 larger by 57%) and more regular (i.e., with Sphericity closer to perfect 1.0), but if the PSD is expressed by volume (mass) fraction, D50 for UA is only 8% higher than that of PAGA. Moreover, the upper values of the PSD (D90) show that the UA batch has fewer large particles than the PAGA batch (also visible in Figure 2 (orange line for UA vs. blue line for PAGA). The additional metric of the powder batches can be a Sauter mean diameter (SMD), which is defined as the diameter of a virtual sphere that has the same volume/surface area ratio for all particles in the batch. This parameter has here practical meaning related to the AM application, because during melting with a laser beam, the heat for melting is transferred mostly through the particle's outer surfaces, while the enthalpy needed for melting is related to particle mass/volume. So more fine powder has a higher area/volume ratio and can be more easily heated with a laser beam (i.e., can be anticipated to have a higher efficiency of thermal conversion). The SMD value is calculated based on PSD and includes the min-max diameter range for not-perfectly spherical particles (i.e., volume and area proportion is obtained for a sum of all ellipsoids in PSD). The batches here, as used in AM trails, have an SMD value of 35.07 microns for gas-atomized (PAGA) powder, and 46.02 microns for ultrasonic-atomized (UA) powder (after classification < 63 microns). This means that the PAGA batch has 31% less area than the UA batch, for the same batch mass. Nevertheless, one can expect that both batches could require similar laser energy density to melt, as the median mass of a particle is not much different, and the weight share of the large particles in the UA batch is smaller, even if the PAGA batch has a higher area for the energy transfer. Additionally, a balanced particle size distribution should positively influence the maximum packing density, and therefore, the generated component's density.

Figure 3 also proves that particularly large or small particles are not included in the UA batch. Therefore, a working hypothesis was that PAGA and UA powders, as compared here, would behave similarly during the manufacturing process, which would be validated by the test campaign.

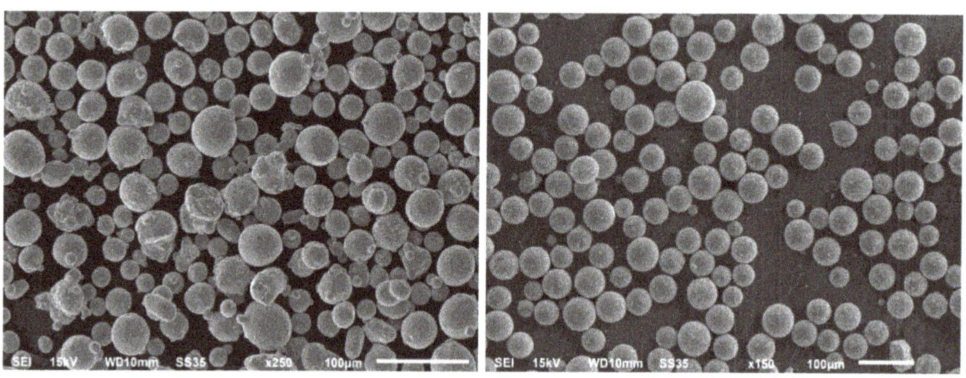

Figure 4. SEM scans of PAGA (**left**) and UA (**right**) particles as tested—the same scale.

Table 2. A. PSD of the powder batches as in the binary image processing (particle counts). B. PSD of the powder batches based on the weight fractions.

	A.	
Production Method	**PAGA**	**UA**
Brand	Carpenter/LPW	3D Lab/ATO
D10 [μm]	16.4	34.1
D50 [μm]	29.4	46.1
D90 [μm]	50.4	55.2
Span = (D90 − D10)/D50	1.078	0.476
Average Sphericity	0.75	0.89
	B.	
Production Method	**PAGA**	**UA**
Brand	Carpenter/LPW	3D Lab—ATO
D10 [μm]	27.3	40.9
D50 [μm]	45.8	49.7
D90 [μm]	73.1	58.2
SMD [μm]	35.1	46.0

3.2. *Physical Properties of the Powders*

The results of the UA vs. PAGA powders flowability properties are shown in Table 3. The secondary metric is a Hausner ratio between tap and apparent densities, which is preferred to be low. This leads to higher flowability and likely reduced porosity after melting.

Table 3. Physical properties of the powder batches.

Production Method	**PAGA**	**UA**
Brand	Carpenter/LPW	3D Lab/ATO
Sieve classification [μm]	15–45	15–60
Tap density [g/cm^3]	4.67	4.72
Apparent density [g/cm^3]	4.29	4.40
Hausner ratio	1.089	1.072
Flow rate [s/50 g]	18.69	14.83

The physical parameters for Carpenter powder after tests here were as typically reported for PAGA powders [21]. Based on measurements, one can conclude that UA powder has better flowability (14.83 s/50 g) and higher density. The UA method is supposed to generate fewer gas bubbles in the powder, because the ATO system operates at low pressure (~1 barA), while the brute force makes PAGA powders of the gas stream at high pressure (30–60 barA). Similar observations for UA powder are reported for other materials, e.g., Titanium Grade 5 [6].

3.3. *Chemical Composition*

The results of the UA powder's chemical composition compared to PAGA powder are listed in Table 4. The variations in the composition were small, and the powder meets the criteria of the AISI 316L specification. More specifically, there was a minor reduction of the manganese (Mn) content from 1.6 to 1.0 %. This change is caused by the minor evaporation of Mn, which is faster than for other elements at atomization temperature, and is resulted in an ATO system from trade-off. The silicon (Si) content in UA powder was slightly higher than the standard AISI 316L specification, but this is the result of the raw material selection, which is a welding wire type 315LSi (here max Si content up to 0.90% per supplier specification). Anyway, the difference in Si content between PAGA and UA batches is negligible here. Another monitored parameter for many powder grades is oxygen (O) content. This value is usually not stated for stainless steel, but a typical 316L powder from gas atomization has 0.05–0.10% of oxygen [21].

Table 4. Chemical composition of PAGA and UA powders.

Element (wt.%)	Specification AISI 316L	Carpenter—LPW CT Powder Range 316L F	3D Lab—ATO System Wire→UA Powder	
Fe	Balanced	Balanced	Balanced	Balanced
C	<0.030	0.027	0.015	0.010
Cr	16.0–18.0	17.8	17.4	17.3
Cu	-	0.02	0.14	0.11
Mn	<2.0	0.98	1.6	1.0
Mo	2.0–3.0	2.31	2.4	2.4
N	<0.10	0.09	0.095	0.040
Ni	10.0–14.0	12.8	10.5	10.5
O	-	0.02	-	0.022
P	<0.045	0.011	0.025	0.025
S	<0.03	0.004	0.007	0.006
Si	<0.75	0.72	0.77	0.76

3.4. Particle Morphology

The particles presented in Figure 4 show a more spherical shape of UA powder from the ATO system (right picture), which is required by AM market, and is indicative of good flowability and a high packing density of the powder (as proven by tests). Furthermore, the SEM picture of UA does not show agglomerates, deformed, and satellite particles, which are common for gas atomization (left picture). The median UA particles are more significant than the PAGA batch due to classification in the ranges of 20–63 µm and 15–45 µm, respectively.

Based on the obtained images, there are visible differences between the two powders. PAGA powder is characterized by a uniform structure with moderately spherical shapes and some satellites on the grains. In the case of UA powder, the particles are more spherical, with some spots visible on their external surface. Some particles have been subjected to deeper analysis to determine the chemical composition in the area of the mentioned spots. The results of the chemical composition analysis of one of the spots are shown in Figure 5. One should note that the surface precipitations have a very small size (<5 µm) and thickness (<2 µm), hence the negligible effect on the melted product properties, while the bulk chemical composition of the batch is fulfilling the AISI 316L specification.

Small brighter spots on the particle's surfaces exhibit increased Cr and Mn content based on the EDS measurements. One hypothesis of the segregation of such elements is related to the lower weight of the mentioned elements compared to Fe and Ni which also have a significant share in the material structure. The EDS analyses were made on UA powder samples to determine if such a phenomenon occurs after the AM process. The results exposed no structural unevenness. Additionally, the powder particles taken from both powder batches were mounted and etched—see Figure 6. It was no registered increased internal porosities in both powder samples. The microstructures of both powder types are characterized by regular grains without any significant imperfections.

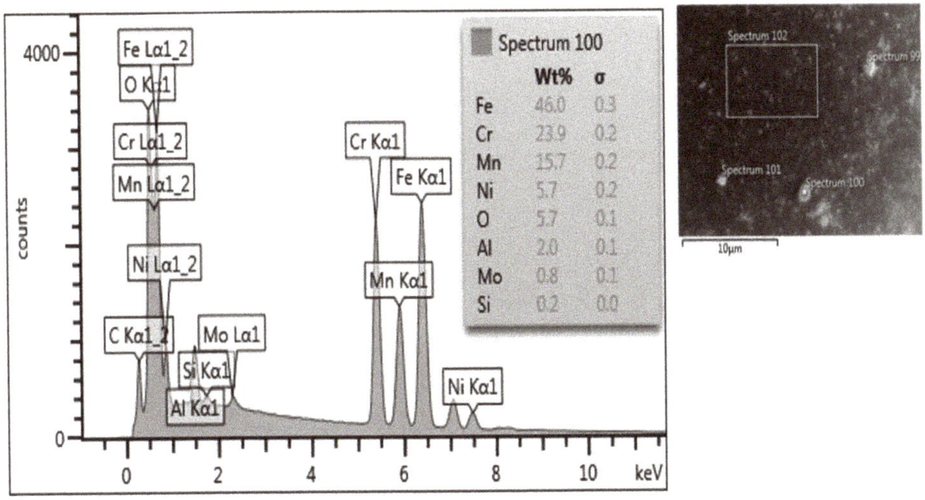

Figure 5. The EDS tests results of one of the spots in the 316L grain surface.

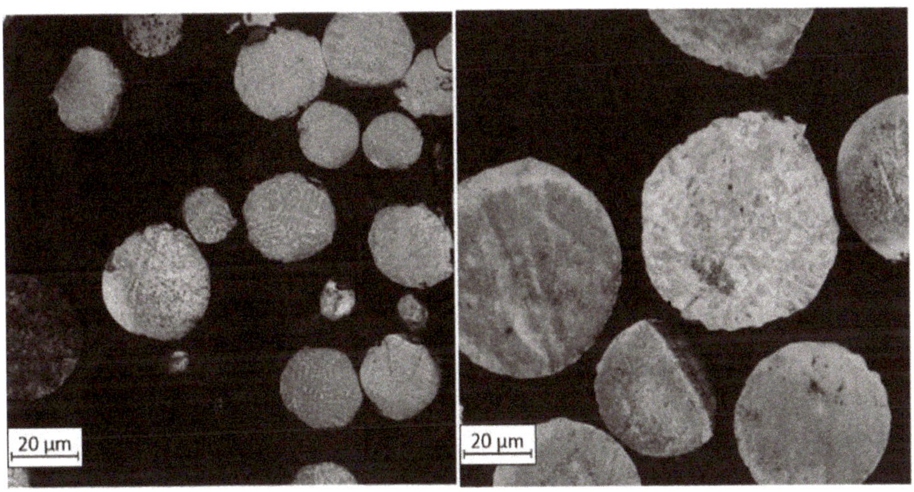

Figure 6. Microstructure of both powders particles: **left**—PAGA; **right**—UA.

3.5. Structure of AM Parts

Thirty-three cubic samples 10 × 10 × 10 mm for each powder batch were additively manufactured with modified parameters, as shown in Table 1. Porosity and defect measurements were taken on the cut surfaces with all layers visible-yz plane (where xy is a substrate plate surface and the z-axis characterized height during the AM process). The obtained results are shown in Figure 7 for porosity vs. the energy density ρ_E and Figure 8 for the maximum defect vs. the energy density ρ_E.

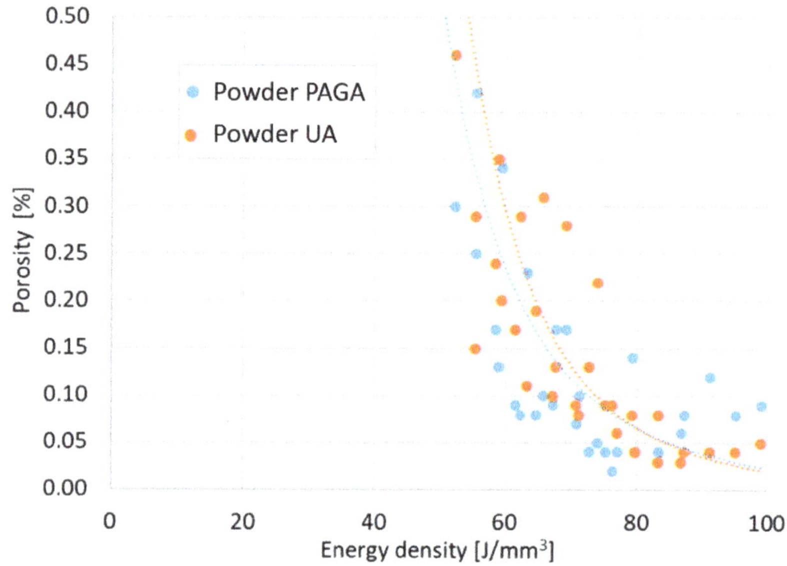

Figure 7. Sample porosity in the function of the laser energy density ρ_E.

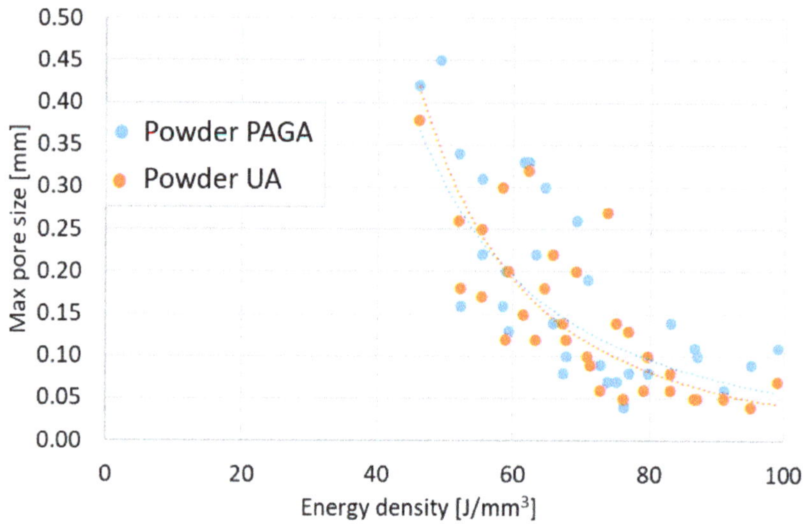

Figure 8. Maximum defect sizes in the function of the laser energy density ρ_E.

Regarding the obtained results, it could be stated that, by using low exposure velocity (700 mm/s), the porosity in samples made of PAGA powder is higher than in the same UA parts manufactured using the same process parameters. After reaching 800 mm/s, the differences in porosity and defect sizes are no different than for typical values of AM process variability. In the case of higher values of exposure velocity (900 mm/s), the porosity increased because of the generation of void related to the "lack of fusion" phenomenon. The primary condition as defined parameters in tensile tests was a minimal porosity value. Based on that rule, the chosen parameters for each powder batch are shown in Table 5. The No. 19 group and No. 22 group were used for the manufacturing process of samples dedicated to tensile testing and impact strength analysis.

Table 5. Parameters selected for tensile samples.

No.	Powder Batch	L_P [W]	e_v [mm/s]	h_d [mm]	ρ_E [J/mm^3]	Porosity [%]	Pore Size [mm]
19	PAGA	220	800	0.12	76.39	0.02	0.04
22	UA	250	800	0.12	86.81	0.03	0.05

The selected printing parameter means that the UA material was melted at almost 15% higher energy density than PAGA (laser power increased by 30 W). This is not surprising if one considers that the PAGA powder has a 31% higher area/volume ratio than the UA batch, as PAGA powder is finer. Because during melting with a laser beam, the heat is transferred through the particle outer surfaces, it is likely to reach higher thermal conversion efficiency for PAGA powder, and is observed to have a lower energy density parameter. However, higher energy doses during melting affect the microstructure in the samples, shown in Figure 9, and strength properties, as discussed later. Some visible differences in the AM parts microstructure are not associated directly with material batch PAGA vs. UA or their composition/microstructure, but rather how intensely the samples were melted with aid of a laser beam [15,22–25]. Most likely the differences in energy density and microstructure shown in Figure 9 could be completely eliminated if UA was either made finer (i.e., at a higher ultrasonic frequency) or classified as finer.

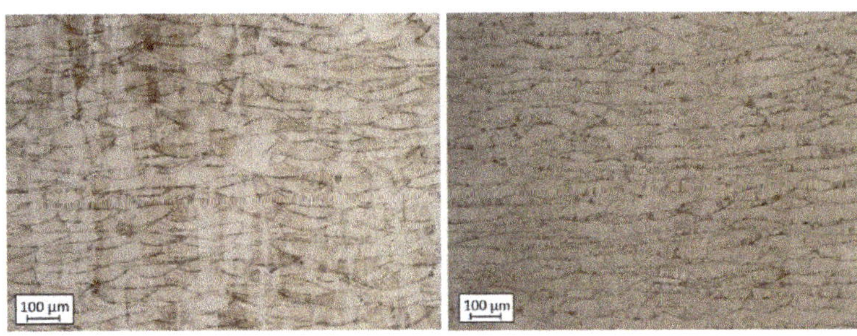

Figure 9. Microstructure of samples manufactured using both powders: **left**—PAGA; **right**—UA.

The additional structure-based test allowed for the determination of material hardness with the HV 0.5 method. The results are shown in Table 6. Samples obtained using UA powder and 8% lower hardness values characterize higher energy density, which is typical behavior during exposure of the powder to the higher energy density in stainless steels. Our previous research [26,27] deliberated such behavior during a deep analysis of the layered structure in AM samples made from the AISI 316L steel. It was proven that very local heat treatment from a laser beam creates a specific cooling effect between consecutive layers (from heat dissipation to deeper layers), which influences the phase changes and the hardness [28], as measured by a sclerometer. The measured thickness of layers in the PAGA samples is smaller than in the same parts produced using UA powder in samples obtained in the current research.

Table 6. Hardness measurements of samples obtained using PAGA and UA powders (#19 and #22 are process parameters from Table 5).

Powder Sample	Value [HV]	Standard Deviation [HV]
PAGA #19	233.67	3.55
UA #22	216.44	3.65

In the case of the AM 316L steel, there is also a phenomenon related to the influence of used higher energy density. Delivering a significant amount of energy into the material's volume causes an increase in the precipitation effect, which affects the drop in the hardness value [13,22,23].

3.6. Tensile Tests with DIC Analysis

The dog-bone-shaped specimens were produced in the L-PBF process according to the ASTM E46696 standard to obtain good quality results, and therefore all specimens were ruptured in the middle of the testing zone. The obtained values of the tensile test are shown in Figure 10, and the most important parameters were set together in Table 7.

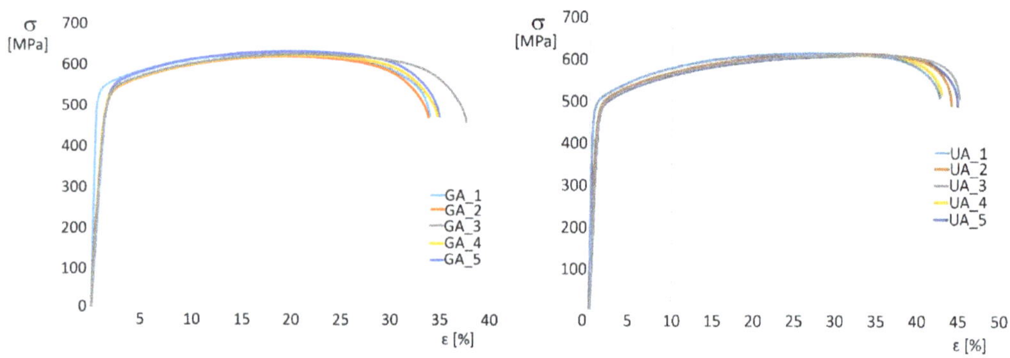

Figure 10. Stress—strain chart of tensile samples made of powders: **left**—PAGA; **righ**—UA.

Table 7. The results registered during tensile testing of samples obtained using PAGA and UA powders.

	PAGA Powder				**UA Powder**			
Sample no.	UTS [MPa]	$R_{0.2}$ [MPa]	E [GPa]	Elong. [%]	UTS [MPa]	$R_{0.2}$ [MPa]	E [GPa]	Elong. [%]
1	621.29	452.65	149.66	34.16	615.91	432.27	167.75	42.96
2	616.65	454.55	150.22	34.03	614.83	437.45	166.22	47.39
3	621.52	452.12	150.44	37.85	612.35	437.42	158.44	46.55
4	618.50	452.27	149.46	34.59	610.56	437.62	162.23	43.09
5	628.65	460.89	149.69	34.31	609.49	437.86	160.47	45.75
Average	621.14	454.78	150.68	35.98	612.68	436.45	163.14	45.41
Standard deviation	4.56	3.78	0.43	1.61	2.58	2.45	3.91	2.03

All samples significantly exceeded the standard AISI 316L specifications, which are:

- ultimate tensile strength UTS > 510 MPa,
- 0.2% offset yield strength R0.2 > 350 MPa,
- elongation at break > 30%,
- Charpy impact resistance KCV > 80 J/cm^2.

One observation from the tensile test was the lack of a visible yield limit point in all samples. Samples made of UA powder represented slightly reduced (by 1.5%) ultimate tensile strength (UTS) and 0.2% offset yield strength ($R_{0.2}$), along with significantly larger (by 9.4% point) elongation at break (column "Elong." in Table 7).

The phenomenon of increased elongation at break makes the UA-powder samples more plastic, but at the same time, they are at a slightly reduced strength compared to the as-built PAGA-powder samples. The Young modulus in UA-based samples is visibly higher (about 10%) compared to the PAGA-based samples. Such a phenomenon was analyzed by Niendorf et al. [29] and Rottger et al. [30], where it was described an influence of used energy density on Young Modulus; it increases with the increase of used energy density. The above-mentioned issues could be better understood while considering the DIC analysis, shown in Figure 11. Such analyses were made to detect any unevenness during the tensile testing process. Additionally, the material behavior at some characteristic stages could be revealed as UTS or yield point ($R_{0.2}$ in Figure 11). In the case of the UA-powder-made samples, a significantly bigger area of the high strain level is visible, compared to the PAGA-powder-made samples. Such results could be useful for identifying some fatigue properties of the material as shear of elastic or plastic areas during the loading process.

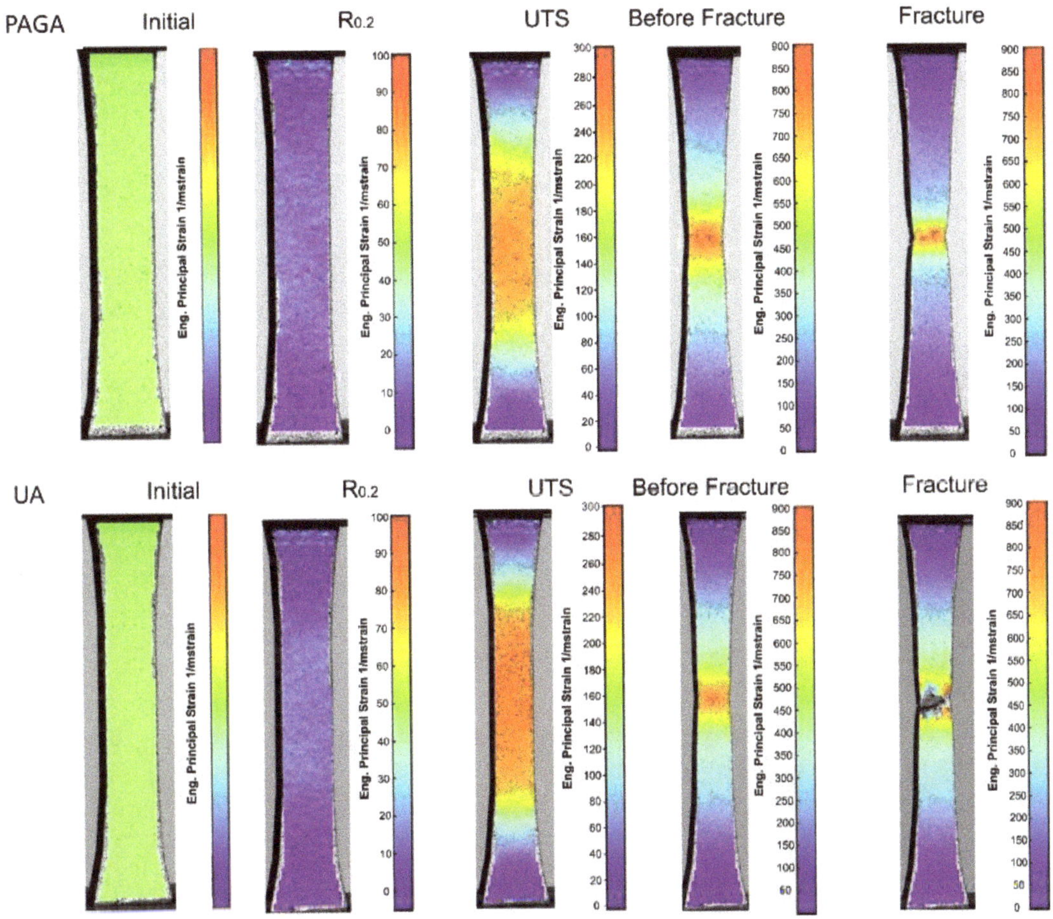

Figure 11. DIC results of samples manufactured using both powders during tensile testing.

DIC analysis revealed different behaviors for each material during strain. Samples made of PAGA powder deformed less equally in a smaller area than samples made of UA until reaching the UTS, where there were visibly increased strain areas in PAGA samples. In both cases, necking is visible, proving the high plasticity of test parts made of PAGA and UA powders.

The difference in tensile properties for samples made of PAGA vs. UA powders can be explained by the fact that, in the case of the UA powder, the selected process parameter (#22 as in Table 5) represented a 15% higher energy density than for PAGA powder (#19 as in Table 5). Previous research by the authors [11,16,27] evaluated this phenomenon for samples made of the same powder (only from PAGA at that time) and adjusted laser parameters. The current and previous [26] test results can be plotted together on one chart (Figure 12) to better observe the effect of the laser energy density on properties like UTS, R0.2, and the elongation at break. The values for UA powder marked on the chart with red marks fall between the other values for larger and smaller energy density collected from other tests for PAGA powder (from the same supplier and test methodology). The general trend observed in all tests is that UTS, $R_{0.2,}$ and elongation values drop after the laser energy increase above 60 J/mm^3. So it is evident that the differences in strength properties between UA-#22 samples and PAGA-#19 samples can be mostly attributed to melting conditions, not to properties of the bulk powder.

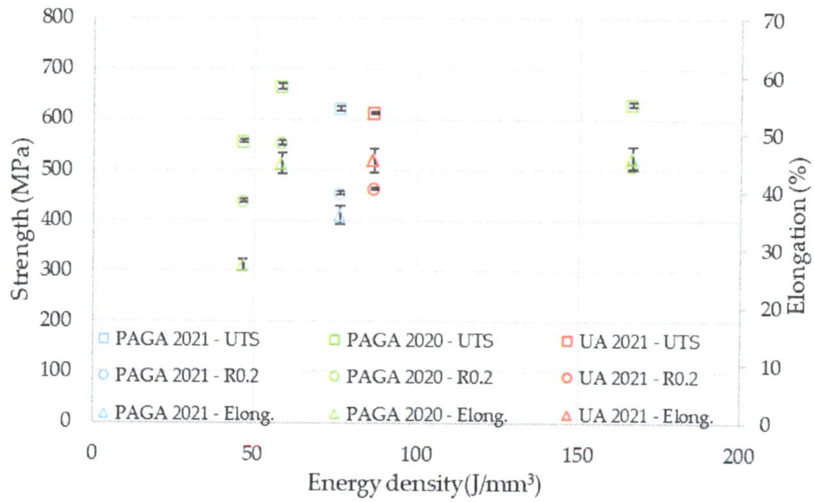

Figure 12. PAGA and UA powder sample strength and elongation vs. laser energy density (data PAGA-2020 from previous own paper [26]).

To better describe two tested materials' behavior, fracture images of selected tensile samples were made (Figure 13).

Figure 13. Fracture images of samples' subjected to tensile testing: (**A**)—PAGA; (**B**)—UA.

Both of the tested samples indicate a mixed character of cracking—mostly plastic, but with local features of brittle-like cracking. Samples made of UA powder indicate more areas of brittle cracking, when at the same time, in the samples made of PAGA powder, there are more voids visible. Such porosity (Figure 13A) could be related to lower energy density used for manufacturing, which caused local lack of fusion between some powder particles.

3.7. Impact Resistance

The last part of the test campaign was a Charpy impact test. As performed in tensile testing, five samples were manufactured using each material. The obtained results are shown on the chart in Figure 14.

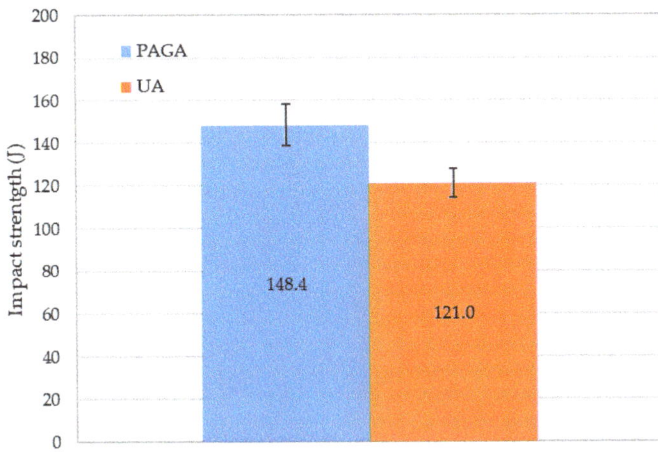

Figure 14. Impact strength of samples manufactured using both powders.

In average values, samples made of UA reached impact energy equal to 121.0 J, whereas at the same time, PAGA samples reached an average of 148.4 J, differing by about 20%. Such a phenomenon could be related to the decreased hardness of the material in samples made of UA powder, which makes the material more plastic.

Also, the effect shown in Figure 12 and described above can be a plausible explanation for impact resistance reduction on UA powder. Nevertheless, all samples easily satisfy the AISI 316L Si specification, which requests that the Charpy impact resistance KCV reach at least 80 J/cm^2 at room temperature.

4. Conclusions

The quality and performance of any 3D printed parts strongly depend on the quality of the powder used in the AM process. However, a preliminary comparison of PAGA vs. UA powder batches allows for the conclusion that UA powder has similar quality as PAGA powder, and can likely be considered to be equivalent. Additionally, it is possible to reach similar AM part properties by properly selecting the production process parameters. Based on the obtained research results, the following conclusions could be drawn:

1. In the case of UA powder, 92.3% of the whole material mass can be classified in the range of 20–63 μm, which is a much bigger value than for the PAGA process (typically 70–75% for steels).
2. The UA material needs almost 15% higher energy density than PAGA (laser power increased by 30 W).
3. Reduced ultimate tensile strength (from 621 MPa to 612 MPa) with higher elongation at break (from 36% to 41%) characterizes samples melted of UA powder utilizing the selected parameters compared with PAGA, which gives obtained values closer to the conventionally made material.
4. The Young's modulus in UA-based samples is visibly higher (about 10%) compared to the PAGA-based samples, which increases with the increase of used energy density.
5. A 20% reduction of the impact strength in the case of UA-made samples was registered, which could be related to the increased plasticity of the material.

Author Contributions: Conceptualization, K.G., M.B. and J.K.; methodology, Ł.S., J.T., L.Ś., M.W., J.Z., Z.K., M.M. and J.Ł.; software, I.S., J.Ł. and M.W.; validation, J.Z., Ł.S., Z.K., J.T. and K.G., formal analysis, L.Ś., M.B., K.G. and J.K.; investigation, Ł.S., J.K., J.Ł., M.B., M.W., Z.K., I.S., J.T. and J.Z.; resources, M.B. and J.K.; data curation, I.S., J.Ł., J.T. and Ł.S.; writing—original draft preparation, J.K., Ł.S., J.T. and M.B.; writing—review and editing, J.T. and J.K.; visualization, I.S. and J.Ł.; supervision, K.G., M.B., J.K. and L.Ś.; project administration, M.B., K.G. and J.K.; funding acquisition, J.K. and L.Ś. All authors have read and agreed to the published version of the manuscript.

Funding: This research was funded by the Military University of Technology, grant number: 22-757/2022.

Institutional Review Board Statement: Not applicable.

Informed Consent Statement: Not applicable.

Data Availability Statement: Not applicable.

Conflicts of Interest: The authors declare no conflict of interest.

References

1. Online Document: ALD Vacuum Technologies, PAGA and VIGA: Metal Powder Inert Gas Atomization Equipment Courte, Vor. Auf RAPID Conf. Exhib. Long Beach 2015. Available online: https://www.ald-vt.com/wp-content/uploads/2017/10/VIGA-EIGA-Brochure.pdf. (accessed on 3 April 2022).
2. Chasoglou, D. Powder Manufacturing & Characterization. Lecture by Hoganas AB for European Powder Metallurgy Association Summer School, June 2016. Available online: https://www.epma.com/document-archive/summer-school-presentations/summer-school-2016/604-powder-manufacturing-characterization/file (accessed on 4 September 2021).
3. Anderson, I. Visualization of Atomization Gas Flow and Melt Break-up Effects in Response to Nozzle Design. In Proceedings of the PowderMet 2011, San Francisco, CA, USA, 18–21 May 2011.
4. Popovich, A.; Sufiiarov, V. Metal Powder Additive Manufacturing. In *New Trends in 3D Printing*; Intech Open Science: London, UK, 2016; Volume 10. Available online: https://www.researchgate.net/publication/305278209_Metal_Powder_Additive_Manufacturing (accessed on 3 September 2021). [CrossRef]
5. Komarov, S.; Kuwabara, M.; Abramov, O.V. High Power Ultrasonics in Pyrometallurgy: Current Status and Recent Development. *ISIJ Int.* **2005**, *45*, 1765–1782. [CrossRef]
6. Bielecki, M.; Kluczyński, J.; Słoboda, Ł. Manufacturing of metallic powders for AM market by ultrasonic atomization method. In Proceedings of the Metal Additive Manufacturing Conference (MAMC 2021), Vienna, Austria, 3–5 November 2021.
7. Rałowicz, R.; Kaczyński, K.; Bielecki, M. A method for Evacuation of Powder Produced by Ultrasonic Atomization and a Device for Implementing This Method. International Patent WO2021009708A1, 21 January 2021.
8. Rałowicz, R.; Rozpędowski, J.; Kaczyński, K.; Żrodowski, L.; Łacisz, W.; Ostrysz, M. Device for the Manufacturing of Spherical Metal Powders by an Ultrasonic Atomization Method. International Patent WO2019092641, 16 May 2019.
9. Fürstenau, J.P.; Wessels, H.; Weißenfels, C.; Wriggers, P. Generating virtual process maps of SLM using powder-scale SPH simulations. *Comput. Part. Mech.* **2020**, *7*, 655–677. [CrossRef]
10. Mitterlehner, M.; Danninger, H.; Gierl-Mayer, C.; Gschiel, H.; Hatzenbichler, M. Processability of Moist Superalloy Powder by SLM. *BHM Berg Hüttenmännische Mon.* **2021**, *166*, 23–32. [CrossRef]
11. Grzelak, K.; Kluczyński, J.; Szachogłuchowicz, I.; Łuszczek, J.; Śnieżek, L.; Torzewski, J. Modification of structural properties using process parameters and surface treatment of monolithic and thin-walled parts obtained by selective laser melting. *Materials* **2020**, *13*, 5662. [CrossRef] [PubMed]
12. Kluczyński, J.; Sniezek, L.; Grzelak, K.; Torzewski, J. The influence of layer re-melting on tensile and fatigue strength of selective laser melted 316L steel. *Intell. Technol. Logist. Mechatron. Syst.* **2018**, *2018*, 115–123.
13. Szachogłuchowicz, I.; Fikus, B.; Grzelak, K.; Kluczyński, J.; Torzewski, J.; Łuszczek, J. Selective Laser Melted M300 Maraging Steel—Material Behaviour during Ballistic Testing. *Materials* **2021**, *14*, 2681. [CrossRef]
14. Blinn, B.; Klein, M.; Gläßner, C.; Smaga, M.; Aurich, J.C.; Beck, T. An investigation of the microstructure and fatigue behavior of additively manufactured AISI 316L stainless steel with regard to the influence of heat treatment. *Metals* **2018**, *8*, 220. [CrossRef]
15. al Mangour, B.; Grzesiak, D.; Yang, J.M. Selective laser melting of TiB2/316L stainless steel composites: The roles of powder preparation and hot isostatic pressing post-treatment. *Powder Technol.* **2017**, *309*, 37–48. [CrossRef]
16. Kluczyński, J.; Śniezek, L.; Grzelak, K.; Oziebło, A.; Perkowski, K.; Torzewski, J.; Szachogłuchowicz, I.; Gocman, K.; Wachowski, M.; Kania, B. Comparison of different heat treatment processes of selective laser melted 316L steel based on analysis of mechanical properties. *Materials* **2020**, *13*, 3805. [CrossRef]
17. Pace, M.L.; Guarnaccio, A.; Dolce, P.; Mollica, D.; Parisi, G.P.; Lettino, A.; Medici, L.; Summa, V.; Ciancio, R.; Santagata, A. 3D additive manufactured 316L components microstructural features and changes induced by working life cycles. *Appl. Surf. Sci.* **2017**, *418*, 437–445. [CrossRef]

18. Stern, F.; Kleinhorst, J.; Tenkamp, J.; Walther, F. Investigation of the anisotropic cyclic damage behavior of selective laser melted AISI 316L stainless steel. *Fatigue Fract. Eng. Mater. Struct.* **2019**, *42*, 2422–2430. [CrossRef]
19. Kluczyński, J.; Śniezek, L.; Grzelak, K.; Torzewski, J.; Szachogłuchowicz, I.; Wachowski, M.; Łuszczek, J. Crack growth behavior of additively manufactured 316L steel-influence of build orientation and heat treatment. *Materials* **2020**, *13*, 3259. [CrossRef]
20. Haferkamp, L.; Haudenschild, L.; Spierings, A.; Wegener, K.; Riener, K.; Ziegelmeier, S.; Leichtfried, G.J. The Influence of Particle Shape, Power Flowability, and Powder Layer Density on Part Density in Laser Powder Bed Fusion. *Metals* **2021**, *11*, 418. [CrossRef]
21. Heiden, M. Pairing 3I6L AM Powder Feedstock with Mechanical Performance: What Really Matters. Available online: https://www.osti.gov/servlets/purl/1643583 (accessed on 3 September 2021).
22. Cheng, B.; Shrestha, S.; Chou, K. Stress and deformation evaluations of scanning strategy effect in selective laser melting. *Addit. Manuf.* **2016**, *12*, 240–251. [CrossRef]
23. Cheng, B.; Chou, K. Geometric consideration of support structures in part overhang fabrications by electron beam additive manufacturing. *CAD Comput. Aided Des.* **2015**, *69*, 102–111. [CrossRef]
24. Heeling, T.; Wegener, K. The effect of multi-beam strategies on selective laser melting of stainless steel 316L. *Addit. Manuf.* **2018**, *22*, 334–342. [CrossRef]
25. Alsalla, H.H.; Smith, C.; Hao, L. Effect of build orientation on the surface quality, microstructure and mechanical properties of selective laser melting 316L stainless steel. *Rapid Prototyp. J.* **2018**, *24*, 9–17. [CrossRef]
26. Kluczyński, J.; Śnieżek, L.; Grzelak, K.; Mierzyński, J. The Influence of Exposure Energy Density on Porosity and Microhardness of the SLM Additive Manufactured Elements. *Materials* **2018**, *11*, 2304. [CrossRef]
27. Kluczyński, J.; Śnieżek, L.; Grzelak, K.; Janiszewski, J.; Płatek, P.; Torzewski, J.; Szachogłuchowicz, I.; Gocman, K. Influence of Selective Laser Melting Technological Parameters on the Mechanical Properties of Additively Manufactured Elements Using 316L Austenitic Steel. *Materials* **2020**, *13*, 1449. [CrossRef]
28. Rahmani, R.; Antonov, M.; Prashanth, K.G. The Impact Resistance of Highly Densified Metal Alloys Manufactured from Gas-Atomized Pre-Alloyed Powders. *Coatings* **2021**, *11*, 216. [CrossRef]
29. Niendorf, A.; Leuders, S.; Riemer, A.; Richard, H.A.; Tröster, T.; Schwarze, D. Highly anisotropic steel processed by selective laser melting. *Metall. Mater. Trans. B* **2013**, *44*, 794–796. [CrossRef]
30. Röttger, A.; Boes, J.; Theisen, W.; Thiele, M.; Esen, C.; Edelmann, A.; Hellmann, R. Microstructure and mechanical properties of 316L austenitic stainless steel processed by different SLM devices. *Int. J. Adv. Manuf. Technol.* **2020**, *108*, 769–783. [CrossRef]

materials

MDPI

Article

Static and Dynamic Mechanical Behaviour of Hybrid-PBF-LB/M-Built and Hot Isostatic Pressed Lattice Structures

David Sommer [1,*], Cemal Esen [2] and Ralf Hellmann [1]

1 Applied Laser and Photonics Group, University of Applied Sciences, Würzburger Straße 45, 63743 Aschaffenburg, Germany
2 Applied Laser Technologies, Ruhr University Bochum, Universitätsstraße 150, 44801 Bochum, Germany
* Correspondence: David.Sommer@th-ab.de

Abstract: We report on a comprehensive study of the mechanical properties of maraging steel body-centred cubic lattice structures fabricated by a hybrid additive manufacturing technology that combines laser powder bed fusion with in situ high-speed milling. As the mechanical properties of additive manufactured components are inferior to, e.g., cast components, surface modifications can improve the mechanical behaviour. Different hybrid additive manufacturing technologies have been designed using additive and subtractive processes, improving process quality. Following this, mechanical testing is performed with respect to static tensile properties and dynamic stress, hardness, and porosity, comparing specimens manufactured by laser powder bed fusion only to those manufactured by the hybrid approach. In addition, the influence of different heat-treatment techniques on the mechanical behaviour of the lattice structures is investigated, namely solution and aging treatment as well as hot isostatic pressing. Thus, the influence of the superior surface quality due to the hybrid approach is evaluated, leading to, e.g., an offset of about 14–16% for the static testing of HIP lattice structures. Furthermore, the dynamic load behaviour can be improved with a finished surface, heading to a shift of the different zones of fatigue behaviour in the testing of hybrid-built specimens.

Keywords: hybrid additive manufacturing; lattice structures; hot isostatic pressing; fatigue behaviour

Citation: Sommer, D.; Esen, C.; Hellmann, R. Static and Dynamic Mechanical Behaviour of Hybrid-PBF-LB/M-Built and Hot Isostatic Pressed Lattice Structures. *Materials* **2023**, *16*, 3556. https://doi.org/10.3390/ma16093556

Academic Editors: Marcin Małek and Janusz Kluczyński

Received: 5 April 2023
Revised: 3 May 2023
Accepted: 4 May 2023
Published: 6 May 2023

1. Introduction

The fundamental freedom of design offered by different additive manufacturing (AM) technologies has stimulated a variety of lightweight structures in mechanical engineering over recent years. These lightweight components are often realized by lattice structures in different forms [1–4]. For instance, periodic structures consisting of unit cells [5–7] or gyroid gradients [8–11] have been utilized for eliminating or relocating material, thus saving weight while maintaining mechanical load capacity [12,13]. Sandwich or infill structures are used for load-dependent design of components for the aerospace industry, modulated for different tensile or compression conditions [14,15]. Based on both experimental studies and simulations, the maximum mechanical load has been correlated to the relative density of different lattice structures [11,16,17].

For metal AM, such as laser powder bed fusion of metals (PBF-LB/M), also referred to as selective laser melting (SLM), the achievable surface accuracy and roughness of components are, in general, inferior to conventional manufacturing methods, limiting applications with demanding requirements in terms of shape and surface conditions. In addition, surface defects and roughness influence mechanical properties such as, e.g., the tensile and fatigue strength. Hence, subtractive surface post-processing is essential for PBF-LB/M components to condition shape and surface properties. This, however, might be challenging for lattice structures due to inaccessible, inner lying areas within the complex lattice arrangement.

In this respect, Zheng et al. and Sarkar et al. investigated the mechanical properties of PBF-LB/M-built components and demonstrated that these are, generally, inferior to conventionally fabricated components [18,19], while surface modifications can improve properties such as, e.g., the fatigue life [20–23].

To circumvent these disadvantages of PBF-LB/M, various hybrid additive manufacturing processes, generally consisting of an additive and a subtractive process, have been developed [24–26]. A very promising hybrid approach is the combination of regular PBF-LB/M as an additive process with high-speed-milling as a subtractive process, capitalizing the advantages of both technologies while evading the disadvantages of the PBF-LB/M-process. In particular, in situ high-speed milling integrated into the powder bed process fosters the capabilities of laser powder bed fusion towards superior surface quality and geometrical accuracy [27–29] with the unique advantage of providing access to component surfaces during the 3D-building phase which might be inaccessible after the component is finished.

In addition, with respect to the mechanical properties of metal structures built by PBF-LB/M, the effects of a post process heat treatment on the microstructure and porosity, hardness, yield and fatigue strength have specifically been addressed [20,30–38]. It has been revealed that a post PBF-LB/M heat treatment increases the mechanical load capacity [39–41]. In particular, solution and aging treatment [42,43] or hot isostatic pressing [44,45] have been proven to reduce porosity and positively modify the microstructure of PBF-LB/M-built parts, in turn leading to improved mechanical properties.

Against this background of optimizing PBF-LB/M of lattice structures for lightweight applications and applying heat treatments to metal AM components, the mechanical properties of body-centred cubic lattice structures are studied. Focussing on the maximum applicable static load and fatigue behaviour upon dynamic load, sole PBF-LB/M-structures are compared with a hybrid process, comprising PBF-LB/M and in-situ 3-axis micro milling. In addition, the effect of different heat-treatments, namely a solution and aging treatment as well as hot isostatic pressing, are investigated by studying porosity, hardness, and microstructure.

2. Materials and Methods

2.1. Hybrid-PBF-LB/M Milling System

Body-centred cubic lattice structures (cf. Section 2.2) are fabricated by additive and hybrid additive manufacturing using a Lumex Avance-25 (Matsuura Machinery GmbH, Wiesbaden, Germany). The machine combines conventional laser powder bed fusion with in situ three-axis high-speed milling, as schematically illustrated in Figure 1 and depicted in Figure 2.

For the milling process, the conventional PBF-LB/M process sequence is paused after several layers (typically ten), allowing access of the milling cutter to all contours of the specimen built thus far. This in turn allows the generation of finished components with superior surface roughness, including the machining of undercuts or internal structures.

The hybrid manufacturing process is conducted under nitrogen atmosphere with less than 3% oxygen level to avoid oxidation, with the build platform being kept at 50 °C to prevent deformation by curling due to thermal stress. The maximum build volume is $250 \times 250 \times 185$ mm^3 in width, depth, and height.

Since the PBF-LB/M-process does not fundamentally differ from industrial standard and this comparative study focuses on the tensile fatigue stress behaviour of the PBF-LB/M- and hybrid-built components, previously optimized and reported process parameters for this machine setup are applied for the manufacturing of the specimens (cf. Table 1) [29]. The machine is equipped with an Ytterbium fibre laser (SPI Lasers plc, Southampton, UK) operating at a wavelength of 1070 µm with a focal diameter of 200 µm.

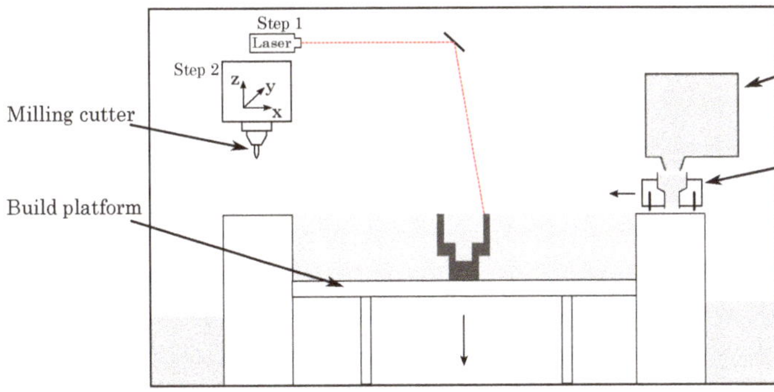

Figure 1. Schematic illustration of the hybrid additive manufacturing unit.

Figure 2. Hybrid additive manufacturing unit, showing the build plate with the recoater and the milling spindle.

Table 1. Laser powder bed fusion parameters.

	Laser Power [W]	Scan Speed [mm/min]	Hatch Distance [µm]
Area	320	700	0.12
Contour	320	1400	-
Support	320	700	0.12

For high-speed milling, the integrated spindle operates at up to 45,000 revolutions per minute and a maximum torque of 1.31 Nm. In addition, a twentyfold tool magazine is available for the change of the milling device during operation. The milling cutters used in this study are solid carbide cutting tools with a nano coating, consisting of aluminium, titanium, and silicone for the reduction of tool wear (Mitsubishi Materials Corporation GmbH, Meerbusch, Germany). The wear characteristics reported in [46] are reinforced by virtue of the dry milling process. Due to the in situ application of the milling process inside the powder bed, no cooling lubricant can be used, leading to elevated temperature conditions and increasing tool wear [47–49]. To analyse the application of the lattice structures, previously determined milling parameters are used, given by Table 2 [29].

As reported before [27,46], the high-speed milling process starts after several built layers, interrupting the PBF-LB/M-process. In our study ten layers with a height of 50 µm each are built before starting the milling process. Within the PBF-LB/M-process, a material allowance of a_t = 150 µm is added to the component structure, being removed gradually by the usage of two different milling cutters, realizing a roughing and a finishing sequence (cf. Figure 3b,c). At first, a_1 = 120 µm of the material allowance are detached by the roughing cutter, machining the surfaces from the last built layers downwards. Afterwards, the finishing cutter removes the remaining a_2 = 30 µm of the material allowance, improving the surface roughness and the final geometrical accuracy.

Table 2. Milling process parameters.

	Z-Pitch [µm]	Spindle Speed [rot/min]	Feed Rate [mm/min]
Roughing Cutter	0.15	30,000	2000
Finishing Cutter	0.1	30,000	1600

Contrary to the roughing cutter, the finishing cutter is working from the bottom to the top of the component as well as the last built layers are spared for the next process cycle. (cf. Figure 3a, where the finishing cutter has left several layers of a_2). This geometrical shift of the milling processes is chosen to avoid thermal distortion as well as to optimize geometrical accuracy and superior surface quality. Due to the PBF-LB/M-process, a heat input is generated, evolving a thermal gradient within the built part. The upper layers exhibit a higher temperature than the layers below, which have already cooled down [50,51]. By virtue of the thermal conditions, a start at the bottom layers working upward is advised and the last built layers should be omitted [52–54].

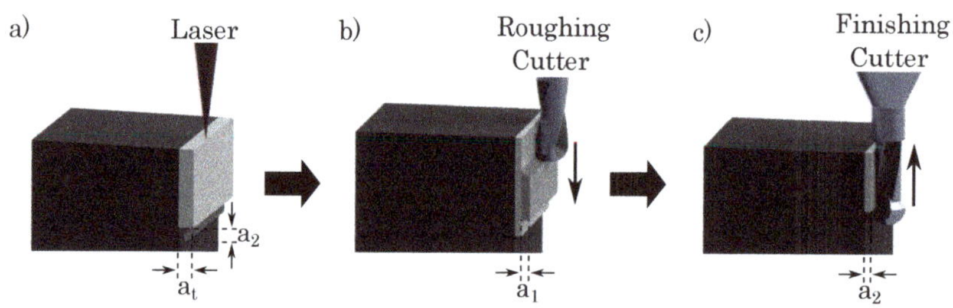

Figure 3. Two stage milling process with two different milling parts for the hybrid system. (**a**) PBF-LB/M process; (**b**) Roughing process; (**c**) Finishing process.

2.2. Specimen Design and Material

To study the effects of an improved surface quality provided by the hybrid process, a body-centred cubic unit cell has been chosen for the lattice structures, as it allows access of the milling cutter and as it represents an often chosen and well-studied structure. The dimensions of the unit cell are 5 × 5 × 5 mm, and the specimens are designed with 3 × 3 × 2 (width × height × depth) cells and a strut diameter of 1 mm, leading to a test specimen with about 13% relative density (cf. Figure 4b). Within the hybrid approach, the manufacturability of the specimens is determined by the three-axis milling system and the diameter of the used milling cutter. In this study, the used milling cutter has a diameter of 2 mm, exhibiting the geometry, shown in Figure 4a). Combined with the three-axis milling system, the milling cutter is machining the specimens on the top and besides (cf. Figure 4b), the undercut surfaces cannot be milled by virtue of the dimensions of the unit cell.

The study is performed processing maraging tool steel X3NiCoMoTi18-9-5 (1.2709, Matsuura Machinery GmbH, Wiesbaden, Germany), a high-nickel steel exhibiting a distinctive mixture of high strength and high hardness, combined with ease of welding. It is preferably employed in the die and tooling industry as well as in structural and aerospace applications [55,56]. The use of maraging steel in PBF-LB/M is also stimulated by the potential of generating cooling channels for the reduction of cycle times in injection moulding [57] of plastics and high-pressure casting of metals [58].

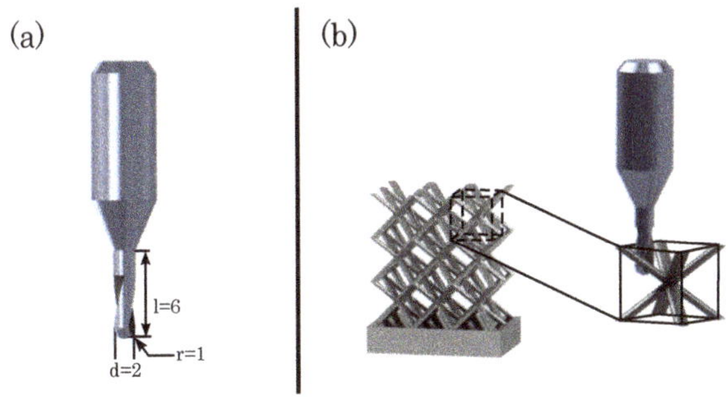

Figure 4. (**a**) Dimensions of the milling cutter, (**b**) milling of lattice structures.

2.3. Post-Processing Procedures

As thermal post-processing of maraging steel, in general, improves mechanical properties of the built parts, the influence of solution treatment combined with aging treatment (SAT) as well as the impact of the hot isostatic pressing (HIP) are studied. For the SAT, a Nabertherm LH 120/12 batch furnace (Nabertherm, Lilienthal, Germany) is employed, enabling heat treatments up to 1200 °C with a maximum heat rate of 10 °C/min. The specimens, completely processed without process gas, are heated to 940 °C at the maximum heat rate, held for 30 min and cooled down to approximately 25 °C with a controlled cooling rate of 10 °C/min (solution treatment). Subsequently, for the aging treatment, the furnace is heated up to 550 °C, held for 4 h and followed by a natural cooling.

The HIP-process is conducted with a Quintus QIH (Quintus Technologies AB, Västeras, Sweden) with a maximum furnace temperature of 1400 °C and a maximum process pressure of 200 MPa. Argon gas is utilized as a process gas. The process starts with a heating up to 1035 °C with a heating rate of 15 °C/min, increasing the pressure to 150 MPa at the same time. This step is followed by a controlled cooling with 10 °C/min down to 940 °C, whereby the previously explained SAT procedure is finished.

The surface properties change within the different heat-treatment procedures (cf. Figure 5). For the SAT, the surface of the specimens shows an effect of blushing, while the HIP specimens obtain a dull appearance. In addition, at the as-built specimens as well as at the HIP specimens, the milling paths are visible, showing a frequent waviness.

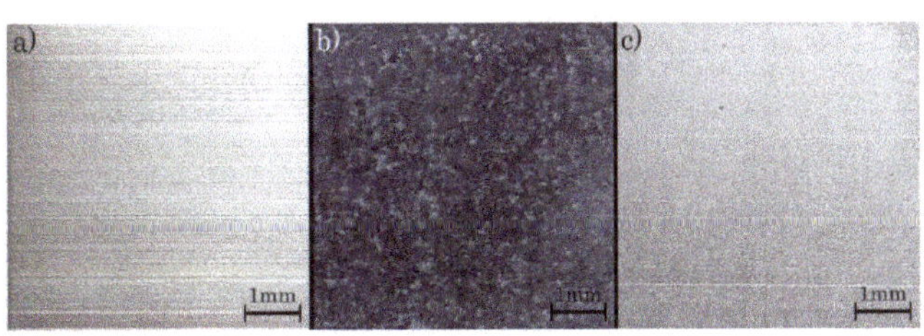

Figure 5. Surfaces of milled specimens: (**a**) as-built, (**b**) SAT and (**c**) HIP.

2.4. Sample Characterization

The mechanical behaviour of the built specimens is tested in the dependence of uniaxial static and dynamic load. For the compression tests with static load, an AG-X plus universal testing machine (Shimadzu, Kyoto, Japan) is used, applying a maximum force of 50 kN due to a precision ball-screw drive. For every state of the heat treatment, machined and PBF-LB/M-built, a batch of three specimens is tested, increasing the statistical significance.

The dynamic testing is executed with a StepLab UD020 (Step Engineering, Resana, Italy) with a maximum compression of 14 kN with a sinusoidal stress-time sequence at a maximum fatigue test frequency of 100 Hz. Analogous to the static tests, three specimens are tested for every level of amplitude, recording the number of cycles until failure.

For the characterization of the microstructure of polished and etched surfaces, a laser scanning microscope VX-2000 (Keyence, Osaka, Japan) is employed. For the examination of the microstructure, the specimens are acid treated with an etching agent according to Adler [59]. Fractographical analysis of cracks appearing during the mechanical testing is conducted with a scanning electron microscope (Maia-3, TESCAN, Dortmund, Germany), using the secondary electron detector (SE) and a SEM voltage of 3 kV.

3. Results and Discussion

In the following, the results of the comparative study between PBF-LB/M-built and hybrid-built lattice structures are shown and discussed. Firstly, mechanical properties with respect to static and the dynamic load for the as built structures are discussed. Secondly, the influence of SAT and HIP is evaluated for both PBF-LB/M-built and hybrid-built lattice structures.

3.1. Static Testing

Upon static load testing, qualitatively the typical performance for maraging steel is observed, revealing a ductile behaviour [60]. In addition, the consecutive collapse of the three individual layers of the built lattice structure is clearly seen in the stress–strain-diagram, until the entire specimen is densely compressed in test direction without entire rupture of the struts, in turn increasing the stress to the maximum applied load, as depicted in Figure 6 [23,33,54].

Comparing the sole PBF-LB/M-built and hybrid-built lattice structures, Figure 5 also shows that the in situ milled lattice structure withstands higher loads, consistently about 14–16% higher, before breaking the single layers than the PBF-LB/M-built structures. The strain and fracture behaviour of PBF-LB/M-built components is significantly influenced by the quantity and dimensions of surface defects [61–63]. Subsequently, the specific energy absorption (SEA) of the hybrid additive manufactured specimens is increased, scaling the applied load to the mass of the specimens [64–66]. While sole PBF-LB/M-built structures show a SEA of about 67.1 kJ/kg, the energy absorption can be raised to about 72.1 kJ/kg due to the surface finish. As shown in the SEM images in Figure 7, the PBF-LB/M-built surface reveals distinct irregularities and defects, associated by adhering powder particles, insufficiently molten regions and superficial cracks, which in turn weaken the mechanical resistance of the specimen. Contrary to this, the in situ machined parts do not show such large defects or cracks on the surface. While the surface roughness, in general, is superior to the PBF-LB/M-built surface roughness, the z-pitch of the milling process can be observed as an irregularity (Figure 7b). The surface quality can be improved to an average surface roughness of $R_a = 1.5–2$ µm, as reported before [46], getting assigned to lead to the superior load capacity of the hybrid-built structures.

After heat treatment with the SAT process, the specimens reveal a distinctively different mechanical behaviour during static testing. As shown in Figure 8, instead of the layer-wise collapse of the different unit cells, the specimens fail completely at a maximum load of 75 MPa (unmachined) and at 81 MPa (machined), exhibiting typical shear fractures [54]. This difference in the mechanical behaviour can be attributed to a change in microstructure as it is expected after the SAT. The martensite (α) phase is dominant for the PBF-LB/M as-built state of maraging steel, indicating only a fractional amount of the austenite (γ) phase [67]. During solution treatment, the γ phase completely merges into the α phase, converting back within aging. The ductile behaviour, observed for the as-built specimen, is significantly reduced and the performance during static load tests is dominated by the austenite phase, leading to a very strong but brittle material [68]. In addition, the hybrid-processed specimens reveal a superior stress–strain response as compared to the sole PBF-LB/M-built parts, attesting the effect of an improved surface condition on the strain and fracture behaviour.

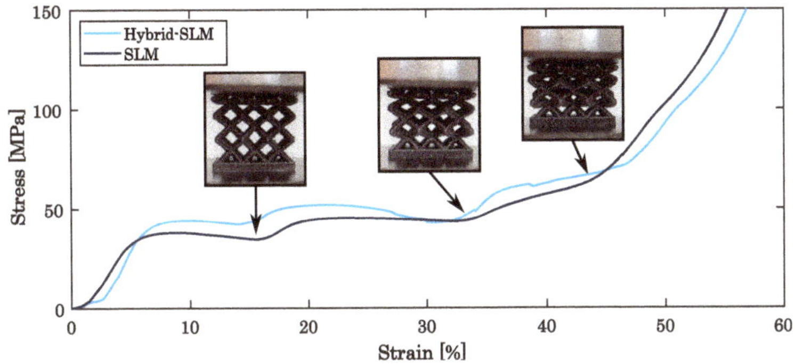

Figure 6. Stress–strain curve of the as-built lattice structures.

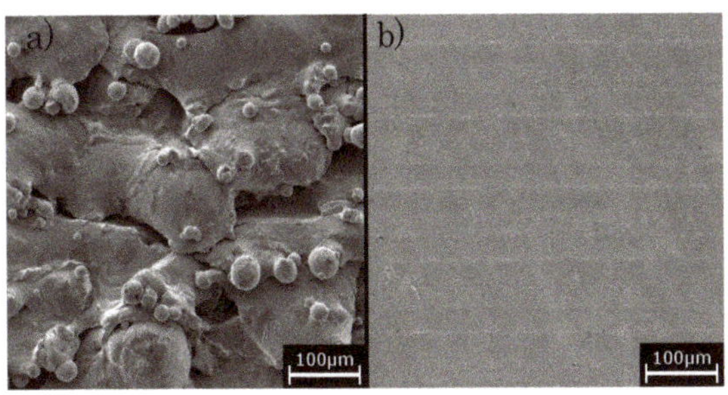

Figure 7. SEM images of (**a**) PBF-LB/M-built surface and (**b**) high-speed milled surface.

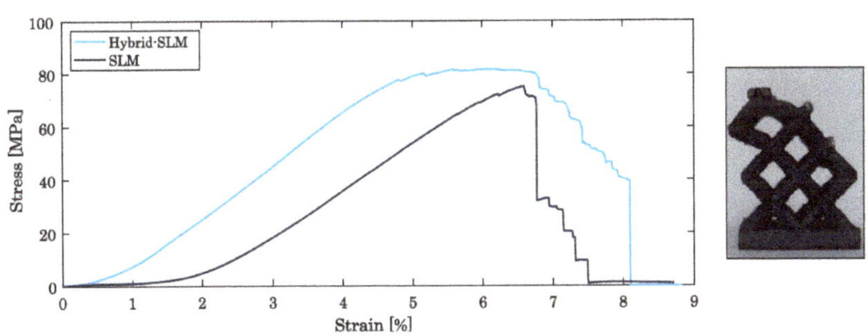

Figure 8. Stress–strain diagram for the comparison of Hybrid PBF-LB/M- and PBF-LB/M-built specimens for SAT lattice structures.

Qualitatively, a similar behaviour is found after HIP of the lattice structures, which fail at a maximum load of 79 MPa (sole PBF-LB/M-built) and 90 MPa (in situ milled), minimizing the ductile material structure as well as the SAT specimens (c.f. Figure 9). The increase in the maximum load of the hybrid-built specimens of about 12% shows an improvement, which is expected to arise from the superior surface quality [69]. In comparison to the SAT specimens, the maximum load for the failure has been increased with about 10 MPa, as reported by previous studies similarly for the comparison of SAT and HIP specimens [70]. The increase of the maximum load can be attributed to the improved material properties due to the HIP process, quantified by the porosity and the hardness of the tested material, measured additionally at cubic test specimens.

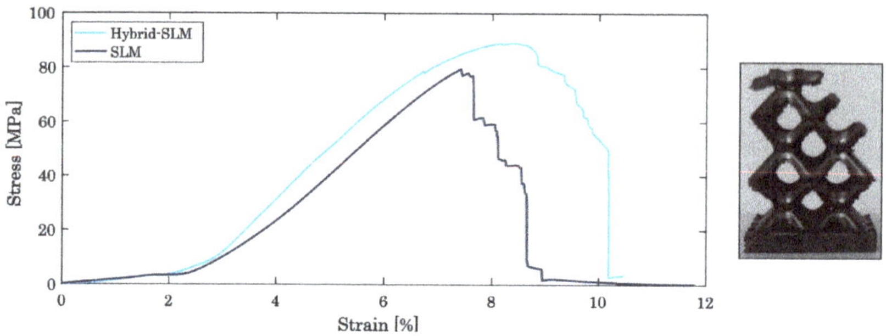

Figure 9. Stress–strain diagram for the comparison of Hybrid PBF-LB/M- and PBF-LB/M-built specimens for HIP lattice structures.

As shown in Figure 10a, the density of post processed components can be raised even though the porosity of as-built components is already high, being about 99.8%. Yet, the density of the SAT-processed specimen reaches about 99.84%, increasing marginally to almost 99.9% in succession of the HIP process. A more pronounced advancement in material properties, however, is achieved for the hardness (c.f. Figure 10b). While the hardness for the PBF-LB/M-built components is approximately at 345 MPa, post-processing improves this characteristic to about 550 MPa for the SAT specimen and to 570 MPa for the HIP specimen.

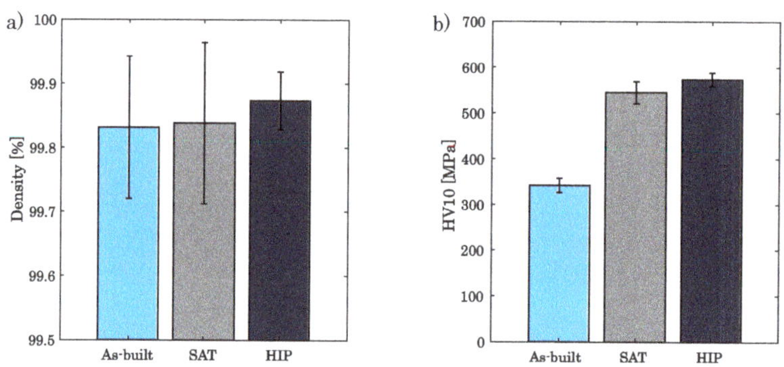

Figure 10. Comparing the as-built, SAT and HIP states for (**a**) density and (**b**) hardness.

While these experimental results indicate an improvement of the mechanical behaviour induced by the heat treatment, they simultaneously lead to a modification of the microstructure of the material [71]. The microstructure of the as-built specimens is unidirectional and dominated by a very fine-grain structure (Figure 11). Due to the different heat treatments, the microstructure changes, leading to a more coarse-grained structure and a dissipated alignment for both heat treatments [60].

Figure 11. Microstructure of the (**a**) as-built, (**b**) SAT and (**c**) HIP specimens.

3.2. Dynamic Testing

Dynamic mechanical testing is performed according to Section 2.4 with a maximum load of 14 kN at a frequency of 50 Hz. For every level of amplitude, three specimens are tested until the final failure occurs, the average being depicted in a Wöhler diagram.

Figure 10 visualises the resulting Wöhler diagrams for the PBF-LB/M-built (Figure 12a) and the hybrid built (Figure 12b) lattice structures, comparing the specimens in the as-built state with the different heat treatment processes. The fatigue behaviour changes in dependence on the cycle number for the various batches of specimens, determining the areas of strength. For both, the PBF-LB/M-built and the hybrid-built lattice structures, the SAT- and HIP-treated specimen reveal a better performance in the low cycle fatigue (LCF) regime, decreasing to assimilable cycle numbers in the section of the high cycle fatigue (HCF). The heat-treated specimens prove a comparable performance for the very high cycle fatigue (VHCF), whereas the as-built structures demonstrate an inferior performance. As for the static load behaviour, the improved hardness and porosity of the heat-treated specimens, especially the HIP specimens, is advantageous for the LCF. With higher cycle numbers, the material performance of the as-built lattice structures is more ductile, raising the VHCF.

For the as-built specimens, the machined structures show a similar overall trend of fatigue behaviour, yet exhibiting an offset with higher cycle numbers for all different sections of strength. This difference can be attributed to the superior surface quality of the machined specimens, as stated before. Figure 13a depicts the crack-analysis for the as-built lattice structures, revealing an endurance failure (zone 1) as a beginning of the crack, including an internal void, possibly enabling the endurance failure. Further, a forced fracture (zone 2) is depicted, causing complete fracture of the specimen. Due to the ductile material structure of the maraging steel in the as-built state, an endurance failure can develop, weakening the structure with every cycle until its complete failure [72].

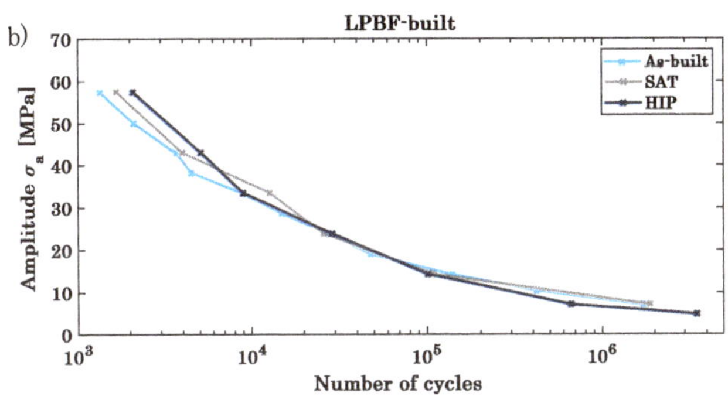

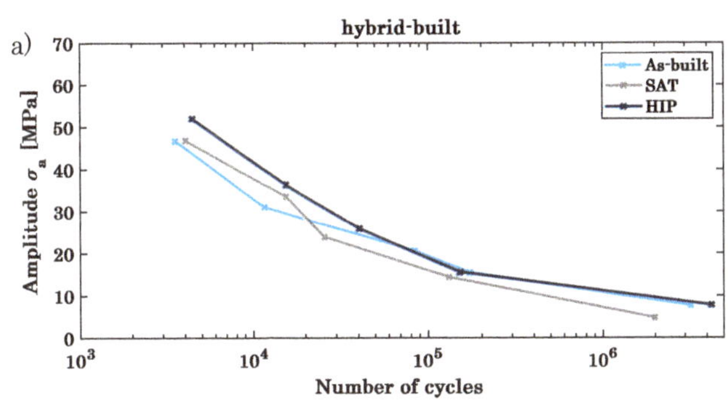

Figure 12. Wöhler diagram for the (**a**) PBF-LB/M-built and (**b**) hybrid-built specimens, each comparing the as-built, SAT and HIP states.

Figure 13. Crack analysis for (**a**) as-built (zone 1: endurance failure, zone 2: forced fracture, getting divided by dashed line), (**b**) SAT and (**c**) HIP specimens for the same applied load.

The SAT structures exhibit almost identical numbers of cycles for the different loads, differing at the beginning of the low cycle fatigue but converging at the following high cycle fatigue, as shown in Figure 14b. Contrary to the as-built specimens, the fracture plain shows a big part of a forced fracture and excludes the endurance failure. The fracture surface only reveals an increased roughness without showing a structure or a development of the fracture (cf. Figure 14b). Due to the heat treatment, the material hardness increased, yet it became more fracturable, leading to a complete failure at a load that excels the maximum capacity.

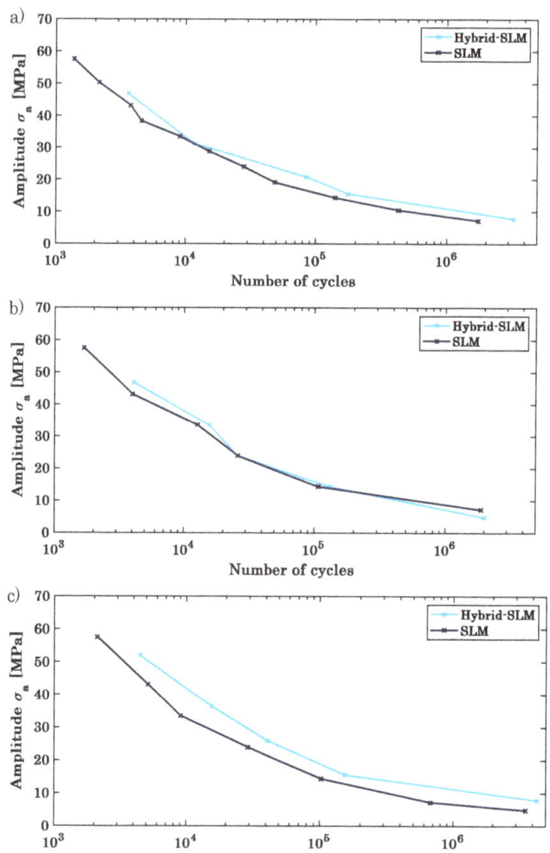

Figure 14. Wöhler diagram for the (**a**) As-built, (**b**) SAT and (**c**) HIP specimens, comparing machined and unmachined specimens.

Similar to the SAT specimens, the HIP lattice structures demonstrate a big part of a forced fracture without any indication of a cyclic developing weakness (cf. Figure 13c). The endurance failure is, for small and high loads, not identifiable at the fracture plain, showing only a little number of porosities, subsequently increasing the maximum load for the lattice structures. Furthermore, the hybrid-built specimens excel the PBF-LB/M-built in the number of the cycles, retaining the offset during all sections of the fatigue testing. The load for the low cycle fatigue is about 10 MPa higher for the machined specimens as for the PBF-LB/M-built as well as the number of cycles in the high cycle fatigue differs about 25,000 at the same applied load (Figure 14).

4. Conclusions

The static and dynamic mechanical behaviour of PBF-LB/M- and hybrid-PBF-LB/M-built lattice structures of maraging tool steel, with the hybrid approach consisting of standard PBF-LB/M combined with in situ high-speed milling, was studied. In addition, the effect of different heat-treatment processes, namely solution treatment combined with an aging treatment process as well as hot isostatic pressing, have been examined, comparing the PBF-LB/M-built and machined specimens. BCC unit cells with a relative density of about 13% were chosen as lattice structures to evaluate the particular influence of the superior surface quality of the in situ milled lattices. For the evaluation of the mechanical load behaviour, static testing has been performed, showing a typical ductile load behaviour for the as-built specimens, while the heat-treated specimens perform with a strong but brittle behaviour, developed due to the post processing.

However, the difference between the PBF-LB/M-built and the hybrid-built specimens persists whether the components are post processes or in the as-built state. As a result of the subsequent machining, the maximum load of the milled specimens exceeds that those generated solely with the PBF-LB/M. The in situ milling can eliminate surface defects and superficial cracks, increasing the mechanical behaviour within the static testing. Further, the machined specimens exceed the PBF-LB/M-built at dynamic testing, as an offset of cycle-numbers is persisting nearly continuously over the different sections of fatigue testing. Partially, milled specimens can be assigned to the VHCF, while the PBF-LB/M-built components fall into the range of HCF at the same load. Overall, this study outlines a comprehensive evaluation of the mechanical properties of hybrid-built lattice structures for static and dynamic testing, highlighting the superior load behaviour for the hybrid approach, combining Laser powder bed fusion and in situ high-speed milling as an innovative manufacturing technology.

Author Contributions: Conceptualization, D.S. and R.H.; methodology, D.S.; validation, D.S. and R.H.; formal analysis, D.S.; investigation, D.S.; resources, R.H.; data curation, D.S.; writing—original draft preparation, D.S.; writing—review and editing, D.S., C.E. and R.H.; visualization, D.S.; supervision, C.E. and R.H.; project administration, R.H.; funding acquisition, R.H. All authors have read and agreed to the published version of the manuscript.

Funding: This research received no external funding.

Data Availability Statement: Not applicable.

Conflicts of Interest: The authors declare no conflict of interest.

References

1. de Pasquale, G.; Luceri, F.; Riccio, M. Experimental Characterization of SLM and EBM Cubic Lattice Structures for Lightweight Applications. *Exp. Mech.* **2019**, *59*, 469–482. [CrossRef]
2. Tao, W.; Leu, M.C. Design of lattice structure for additive manufacturing. In Proceedings of the International Symposium on Flexible Automation, Cleveland, OH, USA, 1–3 August 2016; pp. 325–332. [CrossRef]
3. Xiao, Z.; Yang, Y.; Xiao, R.; Bai, Y.; Song, C.; Wang, D. Evaluation of topology-optimized lattice structures manufactured via selective laser melting. *Mater. Des.* **2018**, *143*, 27–37. [CrossRef]
4. Zhou, H.; Zhang, X.; Zeng, H.; Yang, H.; Lei, H.; Li, X.; Wang, Y. Lightweight structure of a phase-change thermal controller based on lattice cells manufactured by SLM. *Chin. J. Aeronaut.* **2019**, *32*, 1727–1732. [CrossRef]

5. Maskery, I.; Hussey, A.; Panesar, A.; Aremu, A.; Tuck, C.; Ashcroft, I.; Hague, R. An investigation into reinforced and functionally graded lattice structures. *J. Cell. Plast.* **2017**, *53*, 151–165. [CrossRef]
6. Rehme, O.; Emmelmann, C. Rapid manufacturing of lattice structures with selective laser melting. In *Proceedings SPIE 6107, Laser-Based Micropackaging*; SPIE: Bellingham, WA, USA, 2006; Volume 61070K. [CrossRef]
7. Maconachie, T.; Leary, M.; Tran, P.; Harris, J.; Liu, Q.; Lu, G.; Ruan, D.; Faruque, O.; Brandt, M. The effect of topology on the quasi-static and dynamic behaviour of SLM AlSi10Mg lattice structures. *Int. J. Adv. Manuf. Technol.* **2021**, *18*, 4085–4104. [CrossRef]
8. Yan, C.; Hao, L.; Hussein, A.; Young, P. Ti-6Al-4V triply periodic minimal surface structures for bone implants fabricated via selective laser melting. *J. Mech. Behav. Biomed. Mater.* **2015**, *51*, 61–73. [CrossRef]
9. Yang, E.; Leary, M.; Lozanovski, B.; Downing, D.; Mazur, M.; Sarker, A.; Khorasani, A.; Jones, A.; Maconachie, T.; Bateman, S.; et al. Effect of geometry on the mechanical properties of Ti-6Al-4V Gyroid structures fabricated via SLM: A numerical study. *Mater. Des.* **2019**, *184*, 108165. [CrossRef]
10. Ma, S.; Tang, Q.; Feng, Q.; Song, J.; Han, X.; Guo, F. Mechanical behaviours and mass transport properties of bone-mimicking scaffolds consisted of gyroid structures manufactured using selective laser melting. *J. Mech. Behav. Biomed. Mater.* **2019**, *93*, 158–169. [CrossRef]
11. Soro, N.; Saintier, N.; Merzeau, J.; Veidt, M.; Dargusch, M.S. Quasi-static and fatigue properties of graded Ti–6Al–4V lattices produced by Laser Powder Bed Fusion (LPBF). *Addit. Manuf.* **2021**, *37*, 101653. [CrossRef]
12. Maconachie, T.; Leary, M.; Lozanovski, B.; Zhang, X.; Qian, M.; Faruque, O.; Brandt, M. SLM lattice structures: Properties, performance, applications and challenges. *Mater. Des.* **2019**, *183*, 108137. [CrossRef]
13. Yan, C.; Hao, L.; Hussein, A.; Bubb, S.L.; Young, P.; Raymont, D. Evaluation of light-weight AlSi10Mg periodic cellular lattice structures fabricated via direct metal laser sintering. *J. Mater. Process. Technol.* **2014**, *214*, 856–864. [CrossRef]
14. Lippert, R.B.; Lachmayer, R. Bionic inspired infill structures for a light-weight design by using SLM. In Proceedings of the DS 84: DESIGN 2016 14th International Design Conference, Aarhus, Denmark, 15–19 August 2016; pp. 331–340.
15. Smith, M.; Cantwell, W.J.; Guan, Z.; Tsopanos, S.; Theobald, M.D.; Nurick, G.N.; Langdon, G.S. The quasi-static and blast response of steel lattice structures. *J. Sandw. Struct. Mater.* **2011**, *13*, 479–501. [CrossRef]
16. Lei, H.; Li, C.; Meng, J.; Zhou, H.; Liu, Y.; Zhang, X.; Wang, P.; Fang, D. Evaluation of compressive properties of SLM-fabricated multi-layer lattice structures by experimental test and μ-CT-based finite element analysis. *Mater. Des.* **2019**, *169*, 107685. [CrossRef]
17. Mazur, M.; Leary, M.; McMillan, M.; Sun, S.; Shidid, D.; Brandt, M. Mechanical properties of Ti6Al4V and AlSi12Mg lattice structures manufactured by Selective Laser Melting (SLM). In *Laser Additive Manufacturing*; Brandt, M., Ed.; Elsevier: Amsterdam, The Netherlands, 2017; pp. 119–161. [CrossRef]
18. Sarkar, S.; Kumar, C.S.; Nath, A.K. Effects of different surface modifications on the fatigue life of selective laser melted 15–5 PH stainless steel. *Mater. Sci. Eng. A* **2019**, *762*, 138109. [CrossRef]
19. Zheng, M.; Zhang, S.; Xu, J.; Zhang, J.; Hu, Q.; He, H.; Zhao, X. Microstructure and Mechanical Properties of 1.2709 Die Steel by Selective Laser Melting. In *High Performance Structural Materials*; Han, Y., Ed.; Springer: Singapore, 2018; pp. 35–44. [CrossRef]
20. Kan, W.H.; Chiu, L.N.S.; Lim, C.V.S.; Zhu, Y.; Tian, Y.; Jiang, D.; Huang, A. A critical review on the effects of process-induced porosity on the mechanical properties of alloys fabricated by laser powder bed fusion. *J. Mater. Sci.* **2022**, *57*, 9818–9865. [CrossRef]
21. Mahshid, R.; Hansen, H.N.; Højbjerre, K.L. Strength analysis and modeling of cellular lattice structures manufactured using selective laser melting for tooling applications. *Mater. Des.* **2016**, *104*, 276–283. [CrossRef]
22. Mooney, B.; Kourousis, K.I.; Raghavendra, R.; Agius, D. Process phenomena influencing the tensile and anisotropic characteristics of additively manufactured maraging steel. *Mater. Sci. Eng. A* **2019**, *745*, 115–125. [CrossRef]
23. Yang, L.; Yan, C.; Cao, W.; Liu, Z.; Song, B.; Wen, S.; Zhang, C.; Shi, Y.; Yang, S. Compression–compression fatigue behaviour of gyroid-type triply periodic minimal surface porous structures fabricated by selective laser melting. *Acta Mater.* **2019**, *181*, 49–66. [CrossRef]
24. Cortina, M.; Arrizubieta, J.I.; Ruiz, J.E.; Ukar, E.; Lamikiz, A. Latest Developments in Industrial Hybrid Machine Tools that Combine Additive and Subtractive Operations. *Materials* **2018**, *11*, 2583. [CrossRef] [PubMed]
25. Flynn, J.M.; Shokrani, A.; Newman, S.T.; Dhokia, V. Hybrid additive and subtractive machine tools – Research and industrial developments. *Int. J. Mach. Tools Manuf.* **2016**, *101*, 79–101. [CrossRef]
26. Jimenez, A.; Bidare, P.; Hassanin, H.; Tarlochan, F.; Dimov, S.; Essa, K. Powder-based laser hybrid additive manufacturing of metals: A review. *Int. J. Adv. Manuf. Technol.* **2021**, *114*, 63–96. [CrossRef]
27. Sommer, D.; Götzendorfer, B.; Esen, C.; Hellmann, R. Design Rules for Hybrid Additive Manufacturing Combining Selective Laser Melting and Micromilling. *Materials* **2021**, *14*, 5753. [CrossRef] [PubMed]
28. Sarafan, S.; Wanjara, P.; Gholipour, J.; Bernier, F.; Osman, M.; Sikan, F.; Molavi-Zarandi, M.; Soost, J.; Brochu, M. Evaluation of Maraging Steel Produced Using Hybrid Additive/Subtractive Manufacturing. *J. Manuf. Mater. Process.* **2021**, *5*, 107. [CrossRef]
29. Wüst, P.; Edelmann, A.; Hellmann, R. Areal Surface Roughness Optimization of Maraging Steel Parts Produced by Hybrid Additive Manufacturing. *Materials* **2020**, *13*, 418. [CrossRef] [PubMed]
30. Bai, Y.; Wang, D.; Yang, Y.; Wang, H. Effect of heat treatment on the microstructure and mechanical properties of maraging steel by selective laser melting. *Mater. Sci. Eng. A* **2019**, *760*, 105–117. [CrossRef]
31. Uzan, N.E.; Shneck, R.; Yeheskel, O.; Frage, N. Fatigue of AlSi10Mg specimens fabricated by additive manufacturing selective laser melting (AM-SLM). *Mater. Sci. Eng. A* **2017**, *704*, 229–237. [CrossRef]

32. Hanzl, P.; Zetková, I.; Kučerová, L. Structural Changes and Microstructure of Maraging Steel Lattice Structures using Additive Manufacturing. *Manuf. Technol.* **2019**, *19*, 37–41. [CrossRef]
33. Liu, L.; Kamm, P.; García-Moreno, F.; Banhart, J.; Pasini, D. Elastic and failure response of imperfect three-dimensional metallic lattices: The role of geometric defects induced by Selective Laser Melting. *J. Mech. Phys. Solids* **2017**, *107*, 160–184. [CrossRef]
34. Zhang, M.; Yang, Y.; Di, W.; Xiao, Z.; Song, C.; Weng, C. Effect of heat treatment on the microstructure and mechanical properties of Ti6Al4V gradient structures manufactured by selective laser melting. *Mater. Sci. Eng. A* **2018**, *736*, 288–297. [CrossRef]
35. Casalino, G.; Campanelli, S.L.; Contuzzi, N.; Ludovico, A.D. Experimental investigation and statistical optimisation of the selective laser melting process of a maraging steel. *Opt. Laser Technol.* **2015**, *65*, 151–158. [CrossRef]
36. Delgado, J.; Ciurana, J.; Rodríguez, C.A. Influence of process parameters on part quality and mechanical properties for DMLS and SLM with iron-based materials. *Int. J. Adv. Manuf. Technol.* **2012**, *60*, 601–610. [CrossRef]
37. Fortunato, A.; Lulaj, A.; Melkote, S.; Liverani, E.; Ascari, A.; Umbrello, D. Milling of maraging steel components produced by selective laser melting. *Int. J. Adv. Manuf. Technol.* **2018**, *94*, 1895–1902. [CrossRef]
38. Khorasani, A.M.; Gibson, I.; Goldberg, M.; Littlefair, G. A comprehensive study on surface quality in 5-axis milling of SLM Ti-6Al-4V spherical components. *Int. J. Adv. Manuf. Technol.* **2018**, *94*, 3765–3784. [CrossRef]
39. Roudnicka, M.; Misurak, M.; Vojtech, D. Differences in the Response of Additively Manufactured Titanium Alloy to Heat Treatment - Comparison between SLM and EBM. *Manuf. Technol.* **2019**, *19*, 668–673. [CrossRef]
40. Zhao, Z.Y.; Li, L.; Bai, P.K.; Jin, Y.; Wu, L.Y.; Li, J.; Guan, R.G.; Qu, H.Q. The Heat Treatment Influence on the Microstructure and Hardness of TC4 Titanium Alloy Manufactured via Selective Laser Melting. *Materials* **2018**, *11*, 1318. [CrossRef] [PubMed]
41. Gao, M.; Li, L.; Wang, Q.; Ma, Z.; Li, X.; Liu, Z. Integration of Additive Manufacturing in Casting: Advances, Challenges, and Prospects. *Int. J. Precis. Eng.-Manuf.-Green Technol.* **2021**, *9*, 305–322. [CrossRef]
42. Monkova, K.; Zetkova, I.; Kučerová, L.; Zetek, M.; Monka, P.; Daňa, M. Study of 3D printing direction and effects of heat treatment on mechanical properties of MS1 maraging steel. *Arch. Appl. Mech.* **2019**, *89*, 791–804. [CrossRef]
43. Vrancken, B.; Thijs, L.; Kruth, J.P.; van Humbeeck, J. Heat treatment of Ti6Al4V produced by Selective Laser Melting: Microstructure and mechanical properties. *J. Alloys Compd.* **2012**, *541*, 177–185. [CrossRef]
44. Großwendt, F.; Röttger, A.; Strauch, A.; Chehreh, A.; Uhlenwinkel, V.; Fechte-Heinen, R.; Walther, F.; Weber, S.; Theisen, W. Additive manufacturing of a carbon-martensitic hot-work tool steel using a powder mixture–Microstructure, post-processing, mechanical properties. *Mater. Sci. Eng. A* **2021**, *827*, 142038. [CrossRef]
45. Weddeling, A.; Wulbieter, N.; Theisen, W. Densifying and hardening of martensitic steel powders in HIP units providing high cooling rates. *Powder Metall.* **2016**, *59*, 9–19. [CrossRef]
46. Sommer, D.; Pape, D.; Esen, C.; Hellmann, R. Tool Wear and Milling Characteristics for Hybrid Additive Manufacturing Combining Laser Powder Bed Fusion and In Situ High-Speed Milling. *Materials* **2022**, *15*, 1236. [CrossRef] [PubMed]
47. Goindi, G.S.; Sarkar, P. Dry machining: A step towards sustainable machining–Challenges and future directions. *J. Clean. Prod.* **2017**, *165*, 1557–1571. [CrossRef]
48. Xiong, X.; Zhang, H.; Wang, G. Metal direct prototyping by using hybrid plasma deposition and milling. *J. Mater. Process. Technol.* **2009**, *209*, 124–130. [CrossRef]
49. Lopez de Lacalle, L.N.; Angulo, C.; Lamikiz, A.; Sanchez, J.A. Experimental and numerical investigation of the effect of spray cutting fluids in high speed milling. *J. Mater. Process. Technol.* **2006**, *172*, 11–15. [CrossRef]
50. Patterson, A.E.; Messimer, S.L.; Farrington, P.A. Overhanging Features and the SLM/DMLS Residual Stresses Problem: Review and Future Research Need. *Technologies* **2017**, *5*, 15. [CrossRef]
51. Cheng, B.; Shrestha, S.; Chou, K. Stress and deformation evaluations of scanning strategy effect in selective laser melting. *Addit. Manuf.* **2016**, *12*, 240–251. [CrossRef]
52. Kranz, J.; Herzog, D.; Emmelmann, C. Design guidelines for laser additive manufacturing of lightweight structures in TiAl6V4. *J. Laser Appl.* **2015**, *27*, S14001. [CrossRef]
53. Mercelis, P.; Kruth, J.P. Residual stresses in selective laser sintering and selective laser melting. *Rapid Prototyp. J.* **2006**, *12*, 254–265. [CrossRef]
54. Mazur, M.; Leary, M.; Sun, S.; Vcelka, M.; Shidid, D.; Brandt, M. Deformation and failure behaviour of Ti-6Al-4V lattice structures manufactured by selective laser melting (SLM). *Int. J. Adv. Manuf. Technol.* **2016**, *84*, 1391–1411. [CrossRef]
55. ASM International Handbook Committee. *ASM Handbook–Properties and Selection: Irons, Steels, and High-Performance Alloys*; ASM International: Cleveland, OH, USA, 1990.
56. Hall, A.M.; Slunder, C. *The Metallurgy, Behavior, and Application of the 18-Percent Nickel Maraging Steels*; Battelle Memorial Institute: Columbus, OH, USA, 1968.
57. Vojnová, E. The Benefits of a Conforming Cooling Systems the Molds in Injection Moulding Process. *Procedia Eng.* **2016**, *149*, 535–543. [CrossRef]
58. Wiedenegger, A.; Bruckwilder, J.; Deutsch, C. Ecological and Economic Benefits of Additive Manufacturing in High Pressure Die Casting. *Berg-Hüttenmännische Monatshefte* **2021**, *166*, 237–242. [CrossRef]
59. Petzow, G. *Metallographic Etching: Techniques for Metallography, Ceramography, Plastography*, 2nd ed.; Techniques for Metallography, Ceramography, Plastography; ASM International: Almere, The Netherlands, 1999.
60. Bajaj, P.; Hariharan, A.; Kini, A.; Kürnsteiner, P.; Raabe, D.; Jägle, E.A. Steels in additive manufacturing: A review of their microstructure and properties. *Mater. Sci. Eng. A* **2020**, *772*, 138633. [CrossRef]

61. Babamiri, B.B.; Indeck, J.; Demeneghi, G.; Cuadra, J.; Hazeli, K. Quantification of porosity and microstructure and their effect on quasi-static and dynamic behavior of additively manufactured Inconel 718. *Addit. Manuf.* **2020**, *34*, 101380. [CrossRef]
62. Koutiri, I.; Pessard, E.; Peyre, P.; Amlou, O.; de Terris, T. Influence of SLM process parameters on the surface finish, porosity rate and fatigue behavior of as-built Inconel 625 parts. *J. Mater. Process. Technol.* **2018**, *255*, 536–546. [CrossRef]
63. Read, N.; Wang, W.; Essa, K.; Attallah, M.M. Selective laser melting of AlSi10Mg alloy: Process optimisation and mechanical properties development. *Mater. Des.* **2015**, *65*, 417–424. [CrossRef]
64. Zhao, M.; Liu, F.; Fu, G.; Zhang, D.Z.; Zhang, T.; Zhou, H. Improved Mechanical Properties and Energy Absorption of BCC Lattice Structures with Triply Periodic Minimal Surfaces Fabricated by SLM. *Materials* **2018**, *11*, 2411. [CrossRef]
65. Zhang, H.; Zhou, H.; Zhou, Z.; Zeng, H.; Zhang, X.; Yang, J.; Lei, H.; Han, F. Energy absorption diagram characteristic of metallic self-supporting 3D lattices fabricated by additive manufacturing and design method of energy absorption structure. *Int. J. Solids Struct.* **2021**, *226–227*, 111082. [CrossRef]
66. Liu, F.; Zhou, T.; Zhang, T.; Xie, H.; Tang, Y.; Zhang, P. Shell offset enhances mechanical and energy absorption properties of SLM-made lattices with controllable separated voids. *Mater. Des.* **2022**, *217*, 110630. [CrossRef]
67. Jägle, E.A.; Choi, P.P.; van Humbeeck, J.; Raabe, D. Precipitation and austenite reversion behavior of a maraging steel produced by selective laser melting. *J. Mater. Res.* **2014**, *29*, 2072–2079. [CrossRef]
68. Kempen, K.; Yasa, E.; Thijs, L.; Kruth, J.P.; van Humbeeck, J. Microstructure and mechanical properties of Selective Laser Melted 18Ni-300 steel. *Phys. Procedia* **2011**, *12*, 255–263. [CrossRef]
69. Dareh Baghi, A.; Nafisi, S.; Hashemi, R.; Ebendorff-Heidepriem, H.; Ghomashchi, R. Experimental realisation of build orientation effects on the mechanical properties of truly as-built Ti-6Al-4V SLM parts. *J. Manuf. Process.* **2021**, *64*, 140–152. [CrossRef]
70. Guo, X.; Ling, H.; Huang, X. Effect of HIP treatment on the microstructure and mechanical properties of a Ni-based superalloy fabricated by selective laser melted method. *J. Phys. Conf. Ser.* **2020**, *1605*, 012143. [CrossRef]
71. Oliveira, A.R.; Diaz, J.A.A.; Nizes, A.D.C.; Jardini, A.L.; Del Conte, E.G. Investigation of Building Orientation and Aging on Strength–Stiffness Performance of Additively Manufactured Maraging Steel. *J. Mater. Eng. Perform.* **2021**, *30*, 1479–1489. [CrossRef]
72. Mahmoud, D.; Al-Rubaie, K.S.; Elbestawi, M.A. The influence of selective laser melting defects on the fatigue properties of Ti6Al4V porosity graded gyroids for bone implants. *Int. J. Mech. Sci.* **2021**, *193*, 106180. [CrossRef]

materials

Article

Bending Strength of Polyamide-Based Composites Obtained during the Fused Filament Fabrication (FFF) Process

Michał Mazurkiewicz [1], Janusz Kluczyński [1,*], Katarzyna Jasik [1], Bartłomiej Sarzyński [1], Ireneusz Szachogłuchowicz [1], Jakub Łuszczek [1], Janusz Torzewski [1], Lucjan Śnieżek [1], Krzysztof Grzelak [1] and Marcin Małek [2]

1 Institute of Robots & Machine Design, Faculty of Mechanical Engineering, Military University of Technology, 2 Gen. S. Kaliskiego St., 00-908 Warsaw, Poland; michal.mazurkiewicz@student.wat.edu.pl (M.M.); katarzyna.jasik@student.wat.edu.pl (K.J.); bartlomiej.sarzynski@student.wat.edu.pl (B.S.); ireneusz.szachogluchowicz@wat.edu.pl (I.S.); jakub.luszczek@wat.edu.pl (J.Ł.); janusz.torzewski@wat.edu.pl (J.T.); lucjan.sniezek@wat.edu.pl (L.Ś.); krzysztof.grzelak@wat.edu.pl (K.G.)

2 Institute of Civil Engineering, Faculty of Civil Engineering and Geodesy, Military University of Technology, 2 Gen. S. Kaliskiego St., 00-908 Warsaw, Poland; marcin.malek@wat.edu.pl

* Correspondence: janusz.kluczynski@wat.edu.pl

Abstract: The research shows the comparison between two types of polyamide-based (PA) composites and pure, base material. The conducted analysis describes how the additions of carbon fibers and glass microbeads affect the material's properties and its behavior during the bending tests. All samples have been tested in the three main directions available during the FFF process. To extend the scope of the research, additional digital-image-correlation tests and fracture analyses were made. The obtained results indicated a positive influence of the addition of carbon fibers into the material's volume (from 81.39 MPa in the case of pure PA to 243.62 MPa in the case of the PA reinforced by carbon fibers).

Keywords: additive manufacturing; three-point bending; polyamide-based composites; fused filament fabrication

Citation: Mazurkiewicz, M.; Kluczyński, J.; Jasik, K.; Sarzyński, B.; Szachogłuchowicz, I.; Łuszczek, J.; Torzewski, J.; Śnieżek, L.; Grzelak, K.; Małek, M. Bending Strength of Polyamide-Based Composites Obtained during the Fused Filament Fabrication (FFF) Process. *Materials* **2022**, *15*, 5079. https://doi.org/10.3390/ma15145079

Academic Editor: Massimiliano Avalle

Received: 24 June 2022
Accepted: 19 July 2022
Published: 21 July 2022

1. Introduction

Polyamide-based materials, due to their favorable mechanical properties, have the interest of various industrial sectors, including automotive, aviation, and biomedical ones. They are one of the most popular construction materials in many types of industries. PA materials are characterized by relatively high-performance properties such as strength, hardness, stiffness, fatigue, abrasion resistance, and good friction properties because of the low friction coefficient value. Due to their properties, PA materials are increasingly used in additive-manufacturing (AM) technology [1–3]. Parts obtained via AM, made of PA, are characterized by high durability and thermal and chemical resistance; additionally, they can be further processed by the means of subtractive technologies. These materials allow obtaining a good surface quality; hence, they are used not only for the production of prototypes but also as ready-to-use parts: gears, clamps, slide bearings, spacing spools, casings, or bushings [4].

In the case of AM technologies, PA-based materials are broadly used in the fused filament fabrication (FFF) and selective laser sintering (SLS) methods [5–9]. By means of those two methods, it is possible to obtain the geometrically complex parts and nonstandard components, which are characterized by relatively good performance properties and their production is relatively inexpensive in the case of low-volume production [10–14].

The most frequently used types of PA material in the FFF and SLS technologies are PA6, PA11, and PA12 [15–17]. In the case of the SLS technology, PA-made parts have a form of sintered particles, while in the FFF technology, the material is deposited as extruded lines. The connection between powder particles and extruded lines has a form of adhesion joints [18,19]; hence, the semicrystalline character of the PA polymers [20] is better exposed in parts obtained by means of the FFF technology.

An important feature of PA materials that can affect the AM parts' properties is their hygroscopicity (water absorption from the air). Dry polyamide becomes brittle, and its impact strength is worse, while its tensile strength increases. In the opposite situation, the elasticity and impact strength increase, but at the same time, the tensile strength significantly reduces. In practice, when the filament absorbs water (which may constitute approx. 10% of its mass), it affects the adhesion of the part to the substrate plate as well as the adhesion between layers, to finally increase the surface roughness of the outline shells of AM parts. Additionally, such a phenomenon could foster nozzle clogging or a "bubbling effect" caused by overheating the moisture in the filament volume [21]. Another important issue related to PA-based materials is their proneness to warping, distortion, and worse possibility to obtain good-quality shapes in comparison to other available polymers dedicated for AM. Nowadays, with many available devices in the market, this drawback is not as significant as a few years ago, but still, it is a serious challenge, especially in the case of more geometrically complex parts. The basis of this phenomenon is related to the shrinkage stress generated during the crystallization process [22]. The characteristics of the AM technology based on layer-by-layer manufacturing cause an asynchronous volumetric shrinkage, with additional regularity in the arrangement of the molecular chains which, as a result, aggravates the shrinkage of the AM part [23]. The newest trends in PA-based material production for AM lead to weakening the crystallization abilities and hindering the regular arrangement of molecular shapes [24].

From the beginning of polymer manufacturing, researchers have tried to optimize the chemical composition of polymers to obtain the desired properties. An increase in an exact property (i.e., mechanical) worsens other types (electrical or thermal); hence, there are many different modifications of polymers that allow balancing between different material properties. Shirvanimoghaddam et al. [25] reviewed a state-of-the-art research work related to the optimization of polypropylene-based (PP) materials' properties. On the one hand, the authors described the most popular, conventional methods of modification methods (elastomer additions, additional copolymerization, nucleating agents, fillers, etc.).

However, on the other hand, they highlighted some less conventional methods:

- Annealing of the composites as a post-treatment technique;
- The use of ultrasound energy during the processing;
- The potential of advanced fillers such as graphene, CNT, boron nitride, MXene, MoS2, and MOFs;
- Using pretreatment or post-treatment techniques;
- Enhancing the interfacial bonding.

Similarly to PP, the properties of PA-based materials can be selectively altered by using additives or fillers in order to obtain specific material characteristics. The resulting blends allow for improvements in the mechanical, thermal, and electrical properties over the typical properties of the base material. The most common additives are reinforcement fibers (carbon, aramid, and glass), plasticizers, and stabilizers. In many cases, the performance properties of the PA-based composites with fiber addition are greater than the pure, base material [26–28].

On the one hand, Rahim et al. [29] used the PA12 material combined with bioceramic fillers, which improved the tensile strength and stiffness of pure PA12, and on the other hand, Liao et al. [30] revealed the improvement in the mechanical and thermal properties PA 12 polymer reinforced with carbon fibers during the FFF process. Different research paths of the mentioned research [31,32] independently allowed for an improvement in the material properties. Additionally, the authors of [32] proved that the addition of 10% carbon fibers to the PA12 matrix may result in a noticeable increase in bending strength, as well as an increase in other performance parameters. A positive influence of using fibers as an addition to the PA-based materials was proven by Chacón et al. [31] who registered a very good influence of glass fibers in the PA composite samples during the three-point bending tests.

In the available literature, there are many results of some exact PA materials or PA-based composites [32–36]. A kind of approach based on a comparison of different PA materials and AM technologies is also available there. A good example is the work of Türk et al. [37], where the authors characterized the mechanical properties of the obtained samples (Young's modulus, tensile yield point, and flexural-creep modulus). They used samples obtained by the SLS and compared the results with the samples made via the FFF process with the use of ABS. None of the reviewed papers include data on how each addition affects the material properties, which could be helpful to properly describe the influence of an exact addition in comparison to the pure, base material. Hence, this study investigates the influence of using additions (carbon fibers and glass microbeads) in the PA-based material on the strength properties during three-point bending tests. To extend the scope of the research, samples for each material were tested in three different printing directions. Additionally, the fracture mechanism during three-point bending testing was described by means of a macrostructural investigation. Such an analysis would be helpful to determine a proper application of AM parts made of each PA-based material, not only as temporary spare parts, but also as products for the final usage.

2. Materials and Methods

The materials used in this examination included pure polyamide PA6 (PA6 Neat Black) and two composites based on polyamide—PA6 GK10 and PA6 CF15. It has to be mentioned that the word "pure" in the case of PA6 NB denotes a lack of fibers, ceramics particles, etc. The base material is a blend with introduced additions to reduce warpage during the AM process. All materials were delivered by the Spectrum Filaments company (Spectrum Filaments, Pęcice, Poland). PA6 GK10 is a glass-microbead-doped (10% of the total volume of the material) composite. This type of reinforcement improves the temperature resistance and thermal insulation and, furthermore, lowers the influence of humidity on the mechanical properties [38]. PA6 CF15 is a polyamide-based composite which is carbon fiber (CF)-doped (15% of the total volume of the material). The reinforcement lowers hygroscope features and the material shrinkage during the process of extrusion [39] compared to pure PA6. The properties of all studied materials are presented in Table 1.

Table 1. Properties of used materials [38].

Property	PA6 Neat Black	PA6 GK10	PA6 CF15
Density (g/cm^3)	1.25	1.01	1.25
Tensile strength (MPa)	78	87	170
Extension at max. force (%)	4.4	2.7	2
Modulus of elasticity (GPa)	3.4	4.2	15

The FFF technology was used for AM of the samples dedicated for bending tests. The production process was performed by means of a 3DGence Double P255 device (3DGence, Katowice, Poland). All samples of the given material were produced in one process with the additional raft to improve adhesion features. The FDM process production was prepared with the use of 3DGence Slicer 4.0 (3DGence, Katowice, Poland). The FDM process parameters are shown in Table 2.

Table 2. Parameters of the FFF process suggested by the filament supplier.

Material	Filament Diameter (mm)	Nozzle Diameter (mm)	Table Temperature (°C)	Nozzle Temperature (°C)	Infill (%)	Number of Contours
PA6 NB	1.75	0.4	80	260	100	5
PA6GK10	1.75	0.4	80	260	100	5
PA6GK15	1.75	0.4	80	260	100	5

The 3D models of the samples were designed based on the PN-EN ISO 178, by means of CAD software Autodesk Inventor (Autodesk Inc., San Rafael, CA, USA). Apart from testing three different materials, the influence of the samples' orientation during the AM process on the bending strength was also taken into account. Strength tests were performed for five samples for each of the directions: X, Y, and Z. The designation of the axes along which the printing directions were planned is presented in Figure 1. The specimens were investigated in the as-built state.

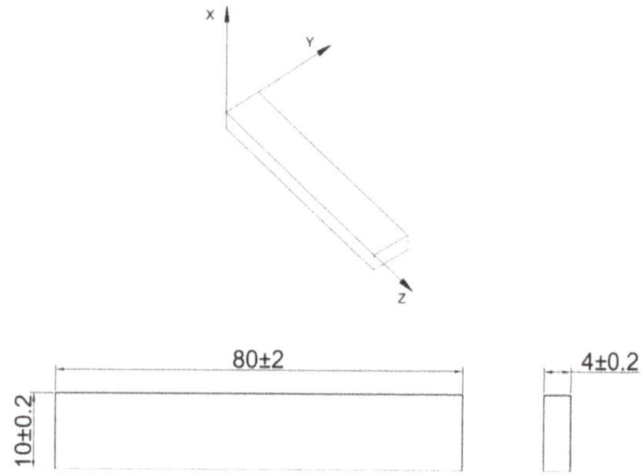

Figure 1. Axes of the test samples and geometric dimensions.

Three-point bending was applied as the main research method. Bending tests were held on the MTS Criterion C4 (MTS Systems GmbH, Berlin, Germany) shown in Figure 1. The research was performed in accordance with ASTM D790-10. The specimen was deformed until its outer surface cracked or until a maximum deformation of 5.0% was reached. The strain rate was 0.01 mm/min (according to procedure A from ASTM D790-10). The specimen was supported on two support points (placed on the longer side of the cross-section) and was loaded with a load lug positioned in the middle of the distance between the supports. Young's modulus, conventional elastic limit ($R_{0.05}$), and conventional yield point ($R_{0.2}$) were calculated in accordance with ISO 178:2019. Furthermore, the digital-image-correlation (DIC) analysis was used to measure the distribution of deformation during the bending test. The measure was performed by means of Dantec Dynamics (Dantec, Ulm, Germany) devices and ISTRA 4D software shown in Figure 2.

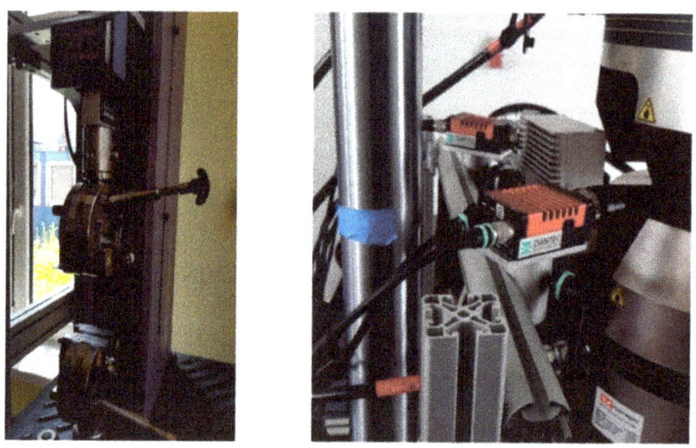

(**a**) (**b**)

Figure 2. (**a**) MTS Criterion C; (**b**) Dantec devices used for the DIC process.

Furthermore, in order to characterize the cracking propagation of the samples during the bending tests, fractographic research was performed. These observations were carried out using the KEYENCE VHX-7000 (Keyence, Osaka, Japan).

3. Results and Discussion

3.1. Flexural Properties—Bending Test

A series of tests were conducted using three different materials, including two composite materials. Samples of each material were tested in three series with different printing directions, as described in Section 2. Table 3 illustrates the results of the flexural strength σ_{fM} for the given samples, and the maximum value of deformation (due to the fact that all samples were cracked brittle, flexural strain at break ε_{fB} is equal to flexure strain at flexural strength ε_{fM}).

Table 3. Registered bending properties of all tested samples.

Material		Average of Flexural Strength σ_{fM} (MPa)	Standard Deviation (Mpa)	Average of Flexural Strain ε_{fB} (%)	Standard Deviation (%)	Young's Modulus E (Gpa)	Standard Deviation (Mpa)
PA6 NB	X	66.75	4.94	4.87	1.32	2.97	0.20
	Y	81.39	1.64	5.25	0.01	1.96	0.45
	Z	23.54	0.70	1.73	0.09	1.11	0.06
PA6 CF15	X	184.18	3.44	3.26	0.18	6.12	0.22
	Y	243.62	10.85	2.53	0.16	11.65	1.41
	Z	8.59	3.65	1.40	0.30	0.92	0.26
PA6 GK10	X	60.10	1.29	3.87	0.21	2.75	0.44
	Y	64.05	0.55	3.95	0.11	0.42	0.34
	Z	16.76	0.44	1.56	0.14	1.09	0.14

The obtained results of the PA6 NB blend are typical for other, available polyamide-based materials acquired during the AM process [40]. The main role of the additions of PA-based blends dedicated for AM is to disturb the crystallization process, which, as a result, reduces the shrinkage stresses. The main principle of such a reduction is increasing the availability of PA molecular chains in a regular arrangement, especially during the cooling process [40]. In the case of the PA6 NB samples, the biggest plasticity of the material was registered, and the greatest strain during three-point bending was obtained for samples oriented along the Y direction (5.25%). The highest flexural strength was obtained for the PA6 CF15 material printed in the Y direction and its average value was 243.62 Mpa. Furthermore, the lowest value was calculated for the same material but produced in the Z direction, and it was 8.59 Mpa. PA6 CF15 was the most brittle material, and the lowest strain was obtained for samples produced in the Z direction (1.40%). Such dispersion of the obtained results in the case of the PA samples with carbon fibers is related to the AM orientation of samples which is additionally affected by the fiber distribution and void fraction. Such a phenomenon was analyzed by Badini et al. [41], where the authors revealed that fibers can be oriented differently inside the AM parts, mostly depending on the orientation in the substrate plated on the AM device. Such a phenomenon is shown in Figure 3.

The highest value of Young's modulus was calculated for the PA6 CF15 material in the Y direction (11.65 Gpa), while the lowest value for PA6 GK10 was produced in the same Y direction (0.42 Gpa). The smallest obtained mechanical-properties values for the PA6 GK10 samples are a result of the nature of glass-microbeads' addition. Their main role is to improve dielectric permittivity, reduce flammability, and increase thermal properties [42]. It is also worth noting that the worse mechanical properties caused by glass microbeads in the polyamide matrix are partially compensated by the enhanced shape and dimensional stability of AM parts [22,25]. Additionally, in the case of the usage of glass microbeads,

Liang et al. [43] made studies on adding glass beads to the polymeric material, which led to an enhancement in Young's modulus and impact strength when, at the same time, the yield strength was reduced as compared to the pure material. Such an enhancement in the case of the impact strength was attributed to the role of glass beads which block crazing at the interface that could not propagate further [25,43].

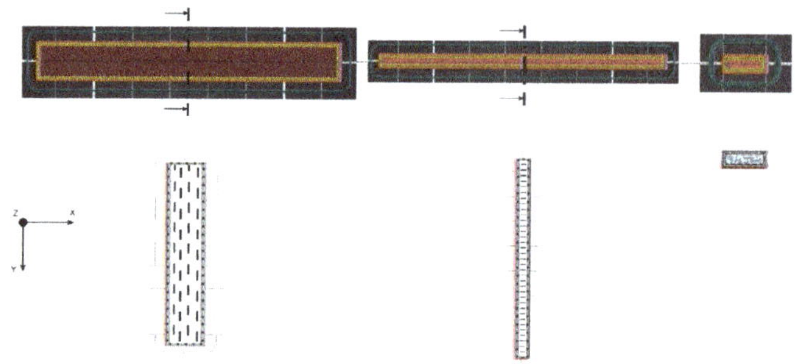

Figure 3. Orientation of fibers inside the tensile samples observed on the cross-section of samples processed according to different orientations, and schematic representation of the expected specimen microstructures [41].

In Figure 4, a representative course of the bending test for the given materials and their orientation is shown.

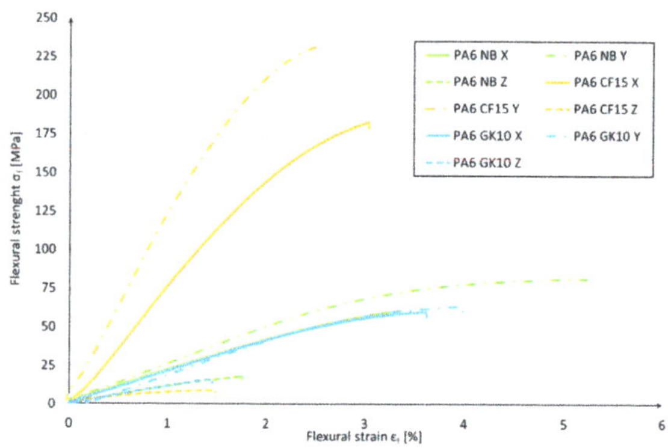

Figure 4. Curves of Flexural Stress (σ_{fM}) versus Flexural Strain (ε_f).

Taking into account the test results from Table 3 and the diagram in Figure 4, comparing with the nonreinforced material (PA6 NB), it can be concluded that the use of carbon fibers positively affected the bending strength of the AM samples. At the same time, in the case of the reinforcement with the use of glass microbeads, the flexural strength significantly decreased. The main reason for the disproportion of mechanical properties in individual production directions is due to the fact that, in the case of the X and Y directions, the load was carried on by the material in the form of contour fibers and filling. Instead, in the case of samples produced in the Z direction, the load was carried on by the connection between the layers, and not by the material itself. These results show that the strength of the material is higher than that of the connection between the layers.

3.2. Digital Image Correlation

During the bending tests, measurements were registered using the DIC method to show the distribution of displacements. Figure 5 shows the measurement results for a maximum strain for each material in different cases of build orientation.

Strain Level in Each Material	DIC Image

PA6NBX—total stain

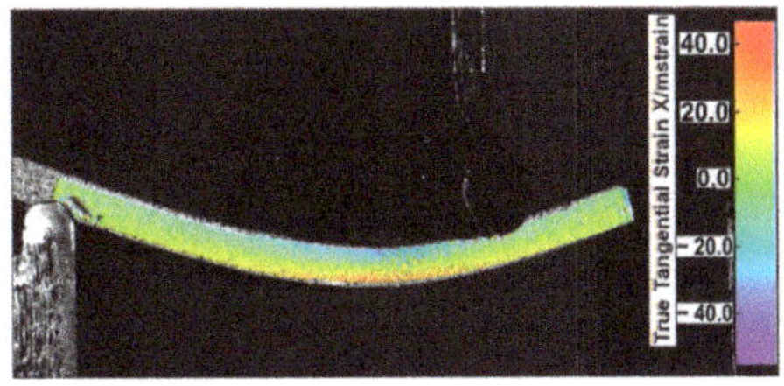

PA6NBY—total stain

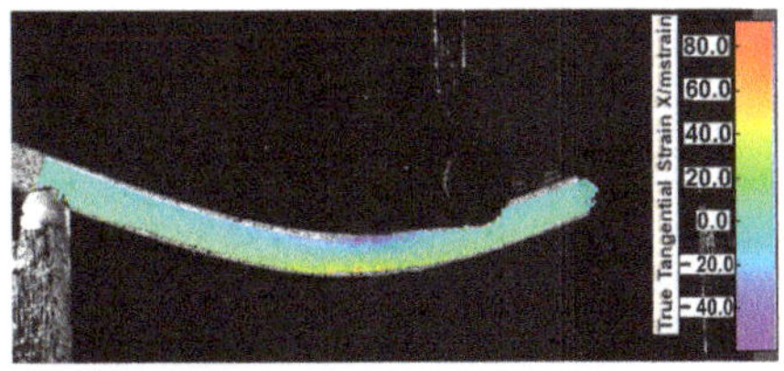

PA6NBZ—total stain

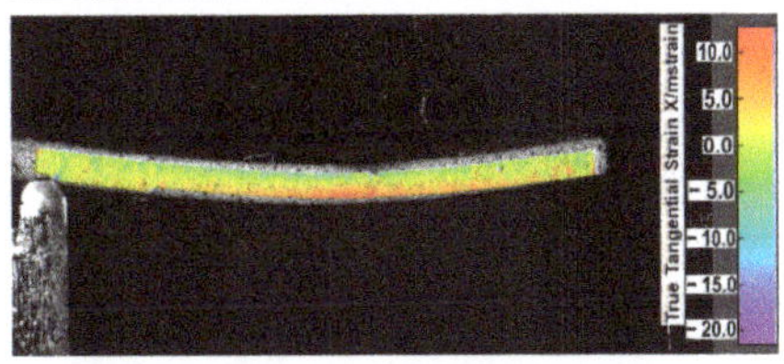

PA6CF15X—total stain

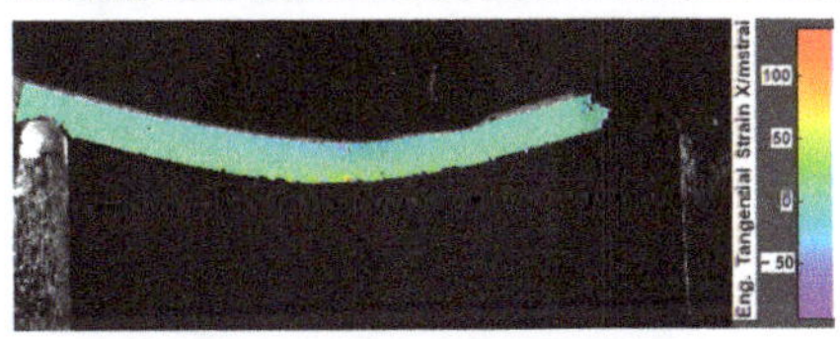

PA6CF15Y—total stain

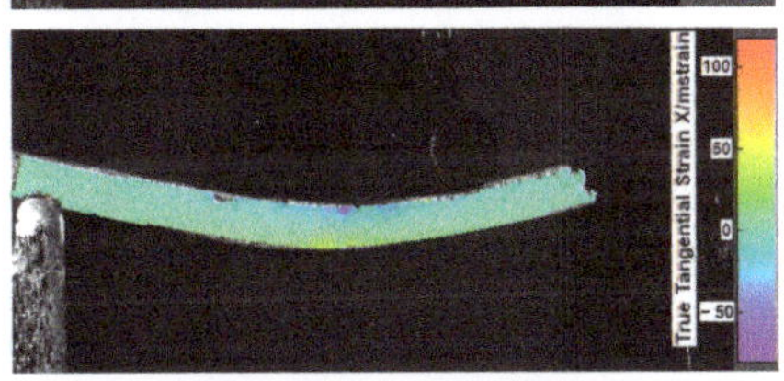

PA6CF15Z—total stain

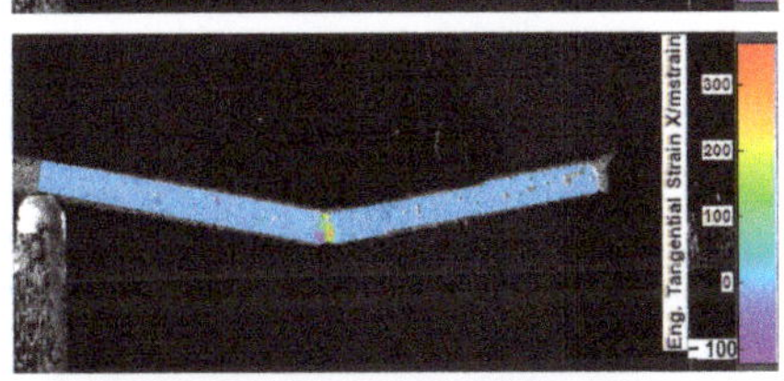

PA6GK10X—total stain

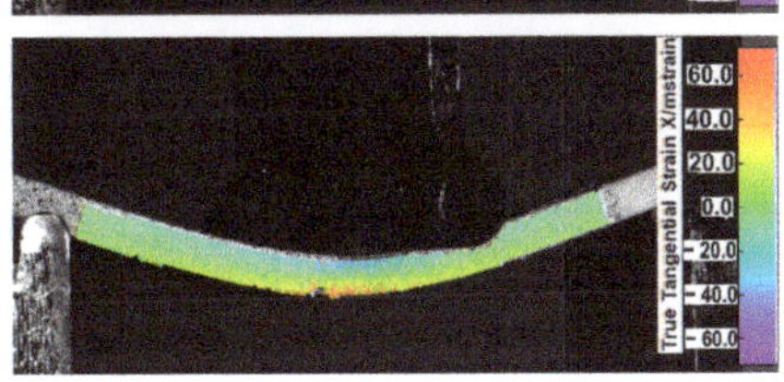

Figure 5. *Cont.*

PA6GK10Y—total stain

PA6GK10Z—total stain

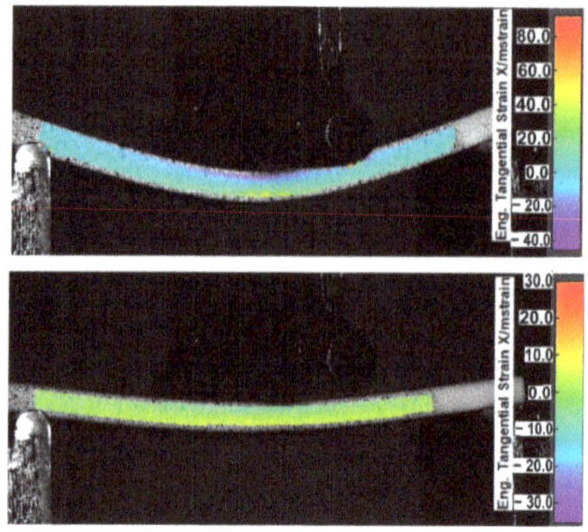

Figure 5. Strain levels registered via the DIC method.

The deformation distribution is uniform in each case in accordance with the theoretical assumptions. On the side of the load handle, deformations with negative values occur, so the material is compressed, while on the opposite side, it is stretched, which is typical in bending testing of conventionally made materials. The most significant differences in the deformation distribution between individual material types are visible for the Z direction. They are caused by loading the material parallel to the direction of joining individual layers and significant differences between the adhesive connection between the layers in each material. The PA6 NB and the PA GK10 materials did not break during the test. On the other hand, the sample made of PA6 CF15 material was fractured, which confirms the negative impact of the reinforcement on the strength of joints between the layers of the material. At the same time, in the case of the PA6 CF15 samples, the highest stiffness of the material was registered. Such a phenomenon could be observed when the fibers and the layer interfaces were parallel to the sample axis. Hence, two orientations of PA6 CF15 were found as the most resistant to bending (X and Y orientations). The principle of this issue is related to the positive influence of fibers on the improvement matrix characteristics (kind of polyamide matrix, matrix crystallinity, and porosity degree). For these reasons, the orientation of a carbon-fiber-reinforced part should be carefully considered when designing an AM process in order to assure proper fiber orientation to obtain the best possible mechanical properties in an exact application.

3.3. *Fracture Analysis*

After the bending tests, fractures of the samples were used for the microscopical examination to extend the research by fracture analysis. The result of a given process of each of the AM PA-based materials in three directions is depicted in Figure 6. In the case of pure PA6 NB samples manufactured along the X direction, a characteristic brittle-like fracture is observed (Figure 6, PA6 NB-X). This phenomenon was mostly caused by the visible voids in the material's structure. In the case of the same material and samples oriented along the Y-axis, the fracture surface is quite similar; however, based on the bending-test results (Figure 5), the total flexural strength is visibly higher. Such behavior could be caused by a significant number of pores. During the material's loading, the cracking mechanism went through the mentioned voids until there was total damage to the sample. In the case of the sample printed in the direction of the Z-axis, the sample was damaged at the point where the subsequent layers of the material were joined, transferring the lowest loads among all three printing directions.

When analyzing samples made of the PA6 GK10 material, similar breakthroughs are visible as in the case of the PA6 NB material samples. Samples made along the Y direction are characterized by a greater number of fracture planes, which confirms the highest bending strength observed during the bending tests. The breakthrough of the sample made along the Z direction looks very similar to the fracture of the corresponding PA6 NB material sample. The crack in the material went directly through the adhesion joint between the subsequent layers.

Figure 6. Fracture images of each sample.

In the last material (PA6 CF15 composite), the fractures reveal the layered material structure and the brittle-like nature of the samples. The fractures of samples made along the X and Y axis look similar, but there are slight differences in the number of pores. Samples arranged along the Y-axis are characterized by a smaller number of pores, which translates into a greater percentage of solid material. Such a phenomenon results in greater material strength and changes the crack generation. Fractures of samples made along the X and Y-axis indicate structural heterogeneity connected with the local stress damming, which promoted a cracking initiation. It is possible to observe that cracking went through several layers, finally making an irregular, multiplanar crack. In the case of samples made along the Z-axis, the fracture has the same nature as two previous materials.

Furthermore, the arrangement of the layer interfaces (related to the orientation on the substrate plate) affects the mechanical behavior of AM parts during the fracture mechanism. That kind of phenomenon is related to the involvement of the debonding mechanism between the stacked layers [30]; it is also an answer for the worst mechanical properties of all samples produced along the Z-axis.

The positive results registered for two orientations (X and Y) are related to the decreased number of pores in the material volume. Such a phenomenon is illustrated in Figure 7, where all visible fibers are significantly surrounded by the base material.

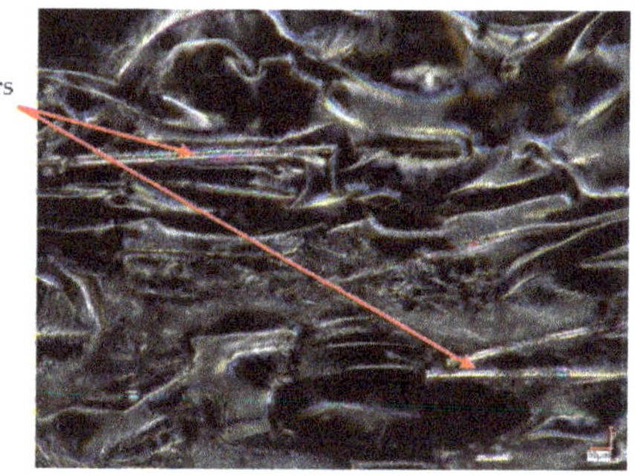

Figure 7. Microscope image of PA6 CF15X structure.

The orientation of the carbon fibers is strictly related to the nozzle movement during each layer fabrication, as was already described and shown in Figure 4. As could be noticed, in the case of the PA6 CF15 samples' fractures, such an orientation of fibers has a significant influence on void generation, which directly affects the mechanical properties. This is why the excellent properties of the PA6 CF15 samples can be explained by the dispersibility and the extrusion inducing a preferential orientation of carbon fiber in the PA matrix [30].

4. Conclusions

The conducted research results allowed us to determine the bending strength of polyamide-based materials, taking into account the influence of additives in the form of carbon fibers and glass microbeads compared to the pure, base material. The analysis of the research results enables the formulation of the following conclusions:

1. The PA6 CF15 material had the highest bending strength in the case of samples produced in the Y direction (243.62 MPa). However, carbon fibers negatively affected the overall ductility of the material.
2. The addition of glass microbeads reduced the bending strength in relation to the PA6 NB base material, regardless of the considered direction of sample manufacturing.
3. Measurements made by the DIC method revealed that the greatest differences in the deformation course took place in the case of the samples produced along the Z direction. They were especially visible in the case of the PA6 CF15 material, where the

carbon fibers negatively affected the quality of the connection between successively deposited layers.
4. The addition of glass balls did not affect the course of material deformation.
5. Fracture analysis showed that the cracking course of the samples produced, irrespective of the direction, had a brittle-like character.
6. The samples produced along the Y-axis had a less complex fracture topography than the samples produced along the X-axis.
7. Regardless of the material type, in the orientation along the Z-axis, the cracks occurred at the joint between the successive layers.

In future research, the authors will try to reduce the presence of voids in the additively manufactured PA-based materials by means of heat treatment in vacuum conditions. A positive result of such postprocessing would be examined from the point of view of mechanical properties.

Author Contributions: Conceptualization, J.K. and M.M. (Michał Mazurkiewicz); methodology, J.K., M.M. (Michał Mazurkiewicz), J.Ł., K.J., I.S. and J.T.; software, M.M. (Marcin Małek); validation, I.S. and K.G.; formal analysis, J.K. and L.Ś.; investigation, M.M. (Michał Mazurkiewicz), J.K., J.Ł., I.S. and J.T.; resources, J.K., M.M. (Michał Mazurkiewicz) and K.G.; data curation, I.S., J.Ł., M.M. (Michał Mazurkiewicz), M.M. (Marcin Małek), K.J. and B.S.; writing—original draft preparation, M.M. (Michał Mazurkiewicz), K.J., B.S., J.K. and J.Ł.; writing—review and editing, J.K. and M.M. (Marcin Małek); visualization, J.K., I.S. and J.Ł.; supervision, J.K.; project administration, J.K.; funding acquisition, J.K. and L.Ś. All authors have read and agreed to the published version of the manuscript.

Funding: This research was funded by the Military University of Technology, grant number: 22-757/2022.

Institutional Review Board Statement: Not applicable.

Informed Consent Statement: Not applicable.

Data Availability Statement: Not applicable.

Conflicts of Interest: The authors declare no conflict of interest.

References

1. Ngo, T.D.; Kashani, A.; Imbalzano, G.; Nguyen, K.T.Q.; Hui, D. Additive manufacturing (3D printing): A review of materials, methods, applications and challenges. *Compos. Part B Eng.* **2018**, *143*, 172–196. [CrossRef]
2. Hamidi, F.; Aslani, F. Additive manufacturing of cementitious composites: Materials, methods, potentials, and challenges. *Constr. Build. Mater.* **2019**, *218*, 582–609. [CrossRef]
3. Praveena, B.A.; Lokesh, N.; Buradi, A.; Santhosh, N.; Praveena, B.L.; Vignesh, R. A comprehensive review of emerging additive manufacturing (3D printing technology): Methods, materials, applications, challenges, trends and future potential. *Mater. Today Proc.* **2022**, *52*, 1309–1313. [CrossRef]
4. Zhu, W.; Yan, C.; Shi, Y.; Wen, S.; Liu, J.; Shi, Y. Investigation into mechanical and microstructural properties of polypropylene manufactured by selective laser sintering in comparison with injection molding counterparts. *Mater. Des.* **2015**, *82*, 37–45. [CrossRef]
5. Sanatgar, R.H.; Campagne, C.; Nierstrasz, V. Investigation of the adhesion properties of direct 3D printing of polymers and nanocomposites on textiles: Effect of FDM printing process parameters. *Appl. Surf. Sci.* **2017**, *403*, 551–563. [CrossRef]
6. Domingo-Espin, M.; Puigoriol-Forcada, J.M.; Garcia-Granada, A.A.; Llumà, J.; Borros, S.; Reyes, G. Mechanical property characterization and simulation of fused deposition modeling Polycarbonate parts. *Mater. Des.* **2015**, *83*, 670–677. [CrossRef]
7. Dul, S.; Fambri, L.; Pegoretti, A. Fused deposition modelling with ABS-graphene nanocomposites. *Compos. Part A Appl. Sci. Manuf.* **2016**, *85*, 181–191. [CrossRef]
8. Giemza, B.; Domański, M.; Deliś, M.; Kapica, D. Tribological properties of 3D printed components. *J. Konbin* **2018**, *48*, 447–463. [CrossRef]
9. Kumar, S.; Kruth, J.P. Composites by rapid prototyping technology. *Mater. Des.* **2010**, *31*, 850–856. [CrossRef]
10. Wang, X.; Jiang, M.; Zhou, Z.; Gou, J.; Hui, D. 3D printing of polymer matrix composites: A review and prospective. *Compos. Part B Eng.* **2017**, *110*, 442–458. [CrossRef]
11. Parandoush, P.; Lin, D. A review on additive manufacturing of polymer-fiber composites. *Compos. Struct.* **2017**, *182*, 36–53. [CrossRef]
12. Melenka, G.W.; Cheung, B.K.O.; Schofield, J.S.; Dawson, M.R.; Carey, J.P. Evaluation and prediction of the tensile properties of continuous fiber-reinforced 3D printed structures. *Compos. Struct.* **2016**, *153*, 866–875. [CrossRef]

13. Syrlybayev, D.; Zharylkassyn, B.; Seisekulova, A.; Akhmetov, M.; Perveen, A.; Talamona, D. Optimisation of strength properties of FDM printed parts—A critical review. *Polymers* **2021**, *13*, 1587. [CrossRef] [PubMed]
14. Vyavahare, S.; Teraiya, S.; Panghal, D.; Kumar, S. Fused deposition modelling: A review. *Rapid Prototyp. J.* **2020**, *26*, 176–201. [CrossRef]
15. Olakanmi, E.O.; Cochrane, R.F.; Dalgarno, K.W. A review on selective laser sintering/melting (SLS/SLM) of aluminium alloy powders: Processing, microstructure, and properties. *Prog. Mater. Sci.* **2015**, *74*, 401–477. [CrossRef]
16. Goodridge, R.D.; Tuck, C.J.; Hague, R.J.M. Laser sintering of polyamides and other polymers. *Prog. Mater. Sci.* **2012**, *57*, 229–267. [CrossRef]
17. Salazar, A.; Rico, A.; Rodríguez, J.; Escudero, J.S.; Seltzer, R.; Martin De La Escalera Cutillas, F. Monotonic loading and fatigue response of a bio-based polyamide PA11 and a petrol-based polyamide PA12 manufactured by selective laser sintering. *Eur. Polym. J.* **2014**, *59*, 36–45. [CrossRef]
18. Jacques, B.; Werth, M.; Merdas, I.; Thominette, F.; Verdu, J. Hydrolytic ageing of polyamide 11. 1. Hydrolysis kinetics in water. *Polymer* **2002**, *43*, 6439–6447. [CrossRef]
19. Rajesh, J.J.; Bijwe, J.; Tewari, U.S.; Venkataraman, B. Erosive wear behavior of various polyamides. *Wear* **2001**, *249*, 702–714. [CrossRef]
20. Tanikella, N.G.; Wittbrodt, B.; Pearce, J.M. Tensile strength of commercial polymer materials for fused filament fabrication 3D printing. *Addit. Manuf.* **2017**, *15*, 40–47. [CrossRef]
21. Costa, S.F.; Duarte, F.M.; Covas, J.A. Thermal conditions affecting heat transfer in FDM/FFE: A contribution towards the numerical modelling of the process: This paper investigates convection, conduction and radiation phenomena in the filament deposition process. *Virtual Phys. Prototyp.* **2015**, *10*, 35–46. [CrossRef]
22. Zhang, X.; Fan, W.; Liu, T. Fused deposition modeling 3D printing of polyamide-based composites and its applications. *Compos. Commun.* **2020**, *21*, 100413. [CrossRef]
23. Abdullah, A.; Rahim, T.; Hamad, W.; Mohamad, D.; Akil, H.; Rajion, Z. Mechanical and cytotoxicity properties of hybrid ceramics filled polyamide 12 filament feedstock for craniofacial bone reconstruction via fused deposition modelling. *Dent. Mater.* **2018**, *34*, e309–e316. [CrossRef] [PubMed]
24. Kumar, R.; Singh, R.; Ahuja, I.P.S.; Karn, K.N. Processing of Melt Flow Compatible Thermoplastic Composites for Solid State Welding Applications. *Mater. Today Proc.* **2019**, *18*, 3167–3173. [CrossRef]
25. Shirvanimoghaddam, K.; Balaji, K.V.; Yadav, R.; Zabihi, O.; Ahmadi, M.; Adetunji, P.; Naebe, M. Balancing the toughness and strength in polypropylene composites. *Compos. Part B Eng.* **2021**, *223*, 109121. [CrossRef]
26. Feng, L.; Wang, Y.; Wei, Q. PA12 powder recycled from SLS for FDM. *Polymers* **2019**, *11*, 727. [CrossRef]
27. Salmoria, G.V.; Leite, J.L.; Paggi, R.A.; Lago, A.; Pires, A.T.N. Selective laser sintering of PA12/HDPE blends: Effect of components on elastic/plastic behavior. *Polym. Test.* **2008**, *27*, 654–659. [CrossRef]
28. Peng, X.; Zhang, M.; Guo, Z.; Sang, L.; Hou, W. Investigation of processing parameters on tensile performance for FDM-printed carbon fiber reinforced polyamide 6 composites. *Compos. Commun.* **2020**, *22*, 100478. [CrossRef]
29. Rahim, T.N.A.T.; Abdullah, A.M.; Akil, H.M.; Mohamad, D.; Rajion, Z.A. The improvement of mechanical and thermal properties of polyamide 12 3D printed parts by fused deposition modelling. *Express Polym. Lett.* **2017**, *11*, 963–982. [CrossRef]
30. Liao, G.; Li, Z.; Cheng, Y.; Xu, D.; Zhu, D.; Jiang, S.; Guo, J.; Chen, X.; Xu, G.; Zhu, Y. Properties of oriented carbon fiber/polyamide 12 composite parts fabricated by fused deposition modeling. *Mater. Des.* **2018**, *139*, 283–292. [CrossRef]
31. Chacón, J.M.; Caminero, M.A.; Núñez, P.J.; García-Plaza, E.; García-Moreno, I.; Reverte, J.M. Additive manufacturing of continuous fibre reinforced thermoplastic composites using fused deposition modelling: Effect of process parameters on mechanical properties. *Compos. Sci. Technol.* **2019**, *181*, 107688. [CrossRef]
32. Salmoria, G.V.; Leite, J.L.; Ahrens, C.H.; Lago, A.; Pires, A.T.N. Rapid manufacturing of PA/HDPE blend specimens by selective laser sintering: Microstructural characterization. *Polym. Test.* **2007**, *26*, 361–368. [CrossRef]
33. Kim, J.; Creasy, T.S. Selective laser sintering characteristics of nylon 6/clay-reinforced nanocomposite. *Polym. Test.* **2004**, *23*, 629–636. [CrossRef]
34. Salmoria, G.V.; Leite, J.L.; Vieira, L.F.; Pires, A.T.N.; Roesler, C.R.M. Mechanical properties of PA6/PA12 blend specimens prepared by selective laser sintering. *Polym. Test.* **2012**, *31*, 411–416. [CrossRef]
35. Caulfield, B.; McHugh, P.E.; Lohfeld, S. Dependence of mechanical properties of polyamide components on build parameters in the SLS process. *J. Mater. Process. Technol.* **2007**, *182*, 477–488. [CrossRef]
36. Liu, Y.; Zhu, L.; Zhou, L.; Li, Y. Microstructure and mechanical properties of reinforced polyamide 12 composites prepared by laser additive manufacturing. *Rapid Prototyp. J.* **2019**, *25*, 1127–1134. [CrossRef]
37. Türk, D.A.; Brenni, F.; Zogg, M.; Meboldt, M. Mechanical characterization of 3D printed polymers for fiber reinforced polymers processing. *Mater. Des.* **2017**, *118*, 256–265. [CrossRef]
38. Spectrumfilaments. PA6 GK10. Available online: https://spectrumfilaments.com/filament/pa6-gk10/(accessed on 9 January 2022).
39. Spectrumfilaments. PA6 CF15. Available online: https://spectrumfilaments.com/filament/pa6-cf15/ (accessed on 9 January 2022).
40. Jia, Y.; He, H.; Peng, X.; Meng, S.; Chen, J.; Geng, Y. Preparation of a new filament based on polyamide-6 for three-dimensional printing. *Polym. Eng. Sci.* **2017**, *12*, 1322–1328. [CrossRef]

41. Badini, C.; Padovano, E.; Camillis, R.D.; Lambertini, V.G.; Pietroluongo, M. Preferred orientation of chopped fibers in polymer-based composites processed by selective laser sintering and fused deposition modeling: Effects on mechanical properties. *J. Appl. Polym. Sci.* **2020**, *137*, 49152. [CrossRef]
42. Isakov, D.; Lei, Q.; Castles, F.; Stevens, C.; Grovenor, C.; Grant, P. 3D printed anisotropic dielectric composite with meta-material features. *Mater. Des.* **2016**, *93*, 423–430. [CrossRef]
43. Liang, J.Z.; Li, R.K.Y. Effect of filler content and surface treatment on the tensile properties of glass-bead-filled polypropylene composites. *Polym. Int.* **2000**, *49*, 170–174. [CrossRef]

materials

MDPI

Article

Performance Properties of Cement–Glass Composite Bricks (CGCB) with Additively Manufactured (AM) Polymeric Scaffolding

Marcin Małek [1], Janusz Kluczyński [2,*], Waldemar Łasica [1], Mateusz Jackowski [1], Ireneusz Szachogłuchowicz [2], Jakub Łuszczek [2], Janusz Torzewski [2] and Krzysztof Grzelak [2]

[1] Institute of Civil Engineering, Faculty of Civil Engineering and Geodesy, Military University of Technology, 2 Gen. S. Kaliskiego St., 00-908 Warsaw, Poland
[2] Institute of Robots & Machine Design, Faculty of Mechanical Engineering, Military University of Technology, 2 Gen. S. Kaliskiego St., 00-908 Warsaw, Poland
* Correspondence: janusz.kluczynski@wat.edu.pl

Abstract: This study provides an alternative to traditional masonry materials: a cement–glass composite brick (CGCB), with a printed polyethylene terephthalate glycol (PET-G) internal scaffolding (gyroidal structure). This newly designed building material consists of 86% waste (78% glass waste, and 8% recycled PET-G). It can respond to the construction market's needs and provide a cheaper alternative to traditional materials. Performed tests showed an improvement in thermal properties after the use of an internal grate in the brick matrix, i.e., an increase in thermal conductivity (5%), and a decrease in thermal diffusivity (8%) and specific heat (10%). The obtained anisotropy of the CGCB's mechanical properties was much lower than the non-scaffolded parts, indicating a very positive effect of using this type of scaffolding in CGCB bricks.

Keywords: additive manufacturing; cement–glass composite bricks; digital image correlation analysis; material extrusion; fused filament fabrication; PET-G; waste disposal

Citation: Małek, M.; Kluczyński, J.; Łasica, W.; Jackowski, M.; Szachogłuchowicz, I.; Łuszczek, J.; Torzewski, J.; Grzelak, K. Performance Properties of Cement–Glass Composite Bricks (CGCB) with Additively Manufactured (AM) Polymeric Scaffolding. *Materials* **2023**, *16*, 1909. https://doi.org/10.3390/ma16051909

Academic Editors: Francesco Baino and Milena Pavlíková

Received: 25 January 2023
Revised: 14 February 2023
Accepted: 23 February 2023
Published: 25 February 2023

1. Introduction

Concrete, as a material, is characterized by high, or very high, compressive strength during static testing. The high compressive strength, significantly increases the brittleness of concrete, with low strain [1]. Conventional steel reinforcement, used in building engineering, is the most popular solution for low material properties during some specified load conditions (i.e., tensile strength at bending). The dramatic growth in the steel price, caused by the COVID-19 pandemic and the war crisis in Ukraine, at the beginning of 2022, has encouraged the development of novel types of concrete–polymer composites [2].

This approach is attractive, since it is possible to additively manufacture the internal reinforcement (called scaffolding). Salazar et al. [3] suggested using a polymeric scaffolding for constructions exposed to four-point bending: cheap, ultra-high-performance concrete (UHPC). The authors [3] considered two types of polymeric reinforcement structures, three-dimensional octets and cubic lattices. A conventional form of concrete reinforcement, using bars, cannot be used in a fully controlled manner, increasing the value of the stochastic factor of placement and orientation of the reinforcing bars. The new concept of design and topological optimization of the internal scaffold for concrete bricks, involves changes to the amount of the material in stress concentration areas [4–8]. The main conclusion of the Salazar et al. [3] work, was the lack of a connection between the material type and the final strength of the concrete–plastic composite. The primary influence on the sample strength was the scaffolding type, not the material type. A different solution to the reinforcement of concrete composites was suggested by Xu et al. [9]. The authors used the finite element method (FEM) to properly prepare a polymeric scaffolding for exact usage in the construction. The main aim of their work was to increase the total strain of the tested samples during three-point bending. The geometry of the samples had been prepared based on the maximum stresses generated during preliminary bending tests. Such a prepared research methodology allowed the elimination of the brittle-like cracking during the sample's fracture. Additionally, a specially designed scaffolding shape and volume, significantly affected the final properties of the manufactured concrete–polymer composites. Another solution, the cement bricks composite (CBC), with the use of an AM polymeric scaffold, was suggested by Qin et al. [10]. The authors used varying geometries of the structure, based on a combination of basic shapes such as hexagons, squares, diamonds, etc. The main aim of their work was similar to [9], i.e., increasing the total strain of the material during bending testing. The authors [10] also used different materials for the AM of the scaffolding: polyamide (PA) filaments and polymeric resins. The final results indicated that PA structures are a much better solution than reinforcement obtained by the AM of resins, especially transparent resins. A quite different approach to using AM in building engineering was shown in Fadeel et al. [11], where the authors produced a new type of form dedicated to brick manufacturing. The mentioned form allowed the production of parts characterized by very complex structures, to increase their total compressive strength. This approach was made possible by using finite element method (FEM) analysis to optimize the final scaffolding geometry, made of ABS material, from the energy dissipation point of view. The CBCB is not the only field of research currently using AM technologies, concrete mortars are also in the mainstream of scientific research. Quite a new approach was shown by Salazar et al. [12], in their work on using an AM spatial mesh made of recycled thermoplastic polymers. This kind of reinforcement provided a two-fold increase in the four-point bending strength. An additional positive side effect of using this kind of reinforcement, is that it ensured better workability of the produced concrete mortar, compared to modified solutions using reinforcing fibers. A different method of using AM in concrete mortars has been demonstrated by Lin et al. [13]. Compared to other technologies, the authors used recycled ABS filaments for the AM processes of reinforcing flat structures. Such structures were deposited with the concrete layer-by-layer, positively affecting the strain–stress characteristics.

In all of the abovementioned technologies using AM in building engineering, researchers have mainly used polymeric scaffolding, with conventional concrete. To increase the number of waste materials in the construction parts, additional elements of the concrete mortar could be used, e.g., waste glass as an aggregate material [14]. The use of PET-G for additional reinforcement is justified by its high chemical resistance [15–17] and good mechanical properties [18,19]. Due to the difference in the materials used for the production of the concrete composite, i.e., the ceramic material and the polymer, the authors noticed that there is a physically justified difference between the thermal conductivities of both materials. As a result of their physical connection, there are no reaction processes between the concrete and the polymer, therefore chemical adhesion does not occur in this case. According to the literature, in such a case, the combination of polymers and ceramic materials leads to the so-called mechanical adhesion, i.e., due to the geometry of the 3D print, the structure was "anchored" in the designed concrete. In such a case, there is a need to check whether, apart from the increase in mechanical parameters, there is no decrease in other parameters that are determinants for concrete in construction, i.e., its insulation. According to the results of the obtained tests, the insulating power of such a composite increased, which reassured the authors of the validity of their work. The increase in insulation was achieved by creating an air layer, which is an ideal insulator, between the contact boundaries of both materials. Despite the significant growth in pro-ecological movements in building engineering, many issues have to be addressed from the point of view of the material's behavior. An additional factor, in the case of this research, is the use of polymeric scaffolding, which needs to be motivated. In the case of a previous study [14], there was a positive effect with the use of scaffolding, but the reinforcing structure did not properly penetrate the scaffold. Therefore, in this study, AM polymeric scaffolding was used to produce a CGCB brick, with waste glass as an aggregate material. The geometry of the scaffold was modified to allow proper penetration of the waste glass parts. In the available research, there is not enough data about using a mixture of waste polymers (as scaffolding) and waste glass (as reinforcing structures). Hence, mechanical and thermal analyses of such materials were made, to determine the basic performance properties of such kinds of composite materials. The suggested approach could be helpful in a further analysis of using waste materials in building engineering.

2. Materials and Methods

2.1. Characteristics of the AM Process and Polymeric Waste Material

Material extrusion (ME) technology was used to produce the internal polymeric scaffolding. The manufacturing process was held on the fused filament fabrication (FFF) device, Prusa Original MK3s (Prusa Research, Prague, the Czech Republic), with the use of PET-G materials, in the form of a 1.75 mm filaments (Spectrum Filaments Ltd., Pecice, Poland). Before the process, the material was dried in the laboratory dryer for 4 h at 65 °C (using the manufacturer's instructions). The processing parameters used were as follows:

- Hotend temperature: 240 °C;
- Substrate plate temperature: 85 °C;
- Layer thickness: 0.2 mm;
- Infill: 100%;
- Part cooling intensity: 40%;
- Printing speed: 50 mm/s;
- Nozzle diameter: 0.4 mm.

As a geometrical scaffolding structure, a gyroidal infill structure was chosen, to ensure an easy flow of the waste glass parts. Such a geometry is available as a default shape of the infill in the PrusaSlicer Software (version 2.4.1), dedicated to the FFF process preparation. To generate a structure as a form of infill, a full monolithic $40 \times 40 \times 160$ mm^3 block was used. In the further steps, the infill amount was set to 5%, and external shells (outline, bottom, and top) were completely removed. The final form of the reinforcing structure is shown in Figure 1.

2.2. Characterization of the Filler: Cement–Glass Mortar

Based on Portland cement (CEM I 42.5R NA [20]), whose specifications were established following EN 196-6:2019-01 [21] and PN EN 196-1:2016-07 [22], tap water, waste glass powder (grain size less than 0.1 mm), and waste glass aggregate (grain size less than 2.0 mm), the cement–glass mortar was proposed as a filler for the CGCB. The composition of the final component's grain sizes (entirely extracted from glass waste packages) was, about 2%, 8%, 14%, 25%, 21%, and 30%, for 0–0.063 mm, 0.063–0.125 mm, 0.125–0.250 mm, 0.250–0.500 mm, 0.500–1.000 mm, and 1.000–2.000 mm, respectively. A well-compacted aggregate pile was achieved by the glass–sand–grain composition, fitted between upper and lower guidelines [23,24]. The glass cullets employed in this investigation had an uneven surface form, which was created by a mechanical or implosive crushing procedure. Additionally, green, brown, and transparent glass particles were employed in this study. Furthermore, a liquid third-generation additive, based on modified polymers, was used to maintain a low water-to-cement ratio (w/c = 0.29), in order to drastically minimize the amount of tap water in the cement–glass mortar filler. Tables 1–3 provide information on the cement and glass waste's chemical make-up and physical characteristics, respectively.

Table 1. Composition of the aggregate and the binder [25].

Compositions		SiO_2	Al_2O_3	Fe_2O_3	CaO	MgO	SO_3	Na_2O	K_2O	TiO_2	Cl
Unit (vol.%)	Cement	19.5	4.9	2.9	63.3	1.3	2.8	0.1	0.9	-	0.05
	Glass	70.0–74.0	0.5–2.0	0.0–0.1	7.0–11.0	3.0–5.0	-	6.0–8.0	7.0–9.0	0.0–0.1	-

Table 2. Properties of the aggregate and the binder [25].

Property	Specific Surface Area [m^2/kg]	Specific Gravity [kg/m^3]	Initial Setting Time [min]	Average Compressive Strength After 28 days [MPa]
Cement	437	3090–3190	176	68.2
Glass	100	2450	-	-

Table 3. CGM mix proportion (1 m^3).

Mix symbol	Cement [kg]	Water [kg]	Chemical Admixture [kg]	Waste Glass Powder [kg]	Waste Glass Aggregate [kg]
CGM	480	140	4.8	117.8	1782.2

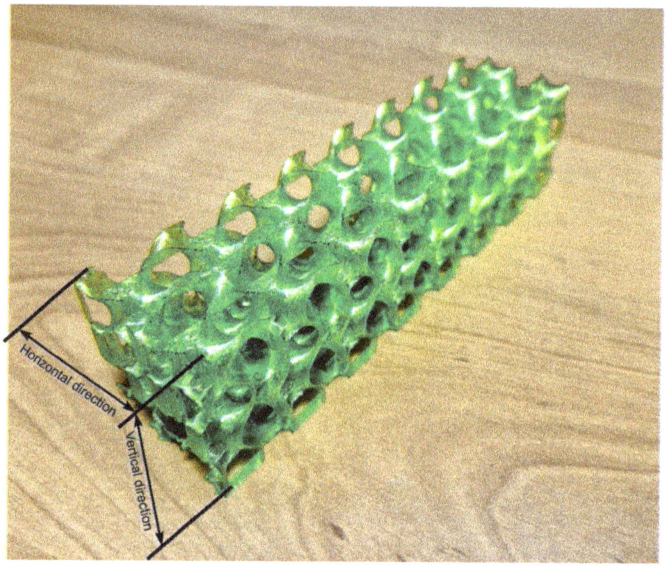

Figure 1. Gyroidal reinforcement structure.

2.3. The Manufacturing Process of Cement–Glass Composite Brick

The formula for the cement–glass mortar was produced using generic, already-in-use techniques for developing high-quality composites [26]. The approach of computational–experimental design was employed. Currently, there are no computational approaches that might guarantee the potential of making high-strength composites with repeatable outcomes without actual experiments, which is why manual stress tests were carried out. General calculations and assumptions were established in the design process, which were later refined after being empirically confirmed (during laboratory tests). The final formula of the CGM [14], used as a filler to create the concrete brick with 3D printed scaffoldings, was, 480 kg of cement, 140 kg of water, 4.8 kg of chemical admixture, 117.8 kg of glass powder, and 1782.2 kg of glass aggregate.

A high-speed planetary mixer, with three ranges of stirrer rotation speed, was used to combine all the dry ingredients mentioned above, for one minute. After adding the wet components, the mixing procedure lasted for an additional four minutes. A medium speed was selected, to thoroughly combine all ingredients (second of three possible rates). The CGM was then compressed on a vibrating table, after being put into molds with previously printed scaffolding. The vibration duration for a single layer was about 30 s. Then, using a regular knife, dampened with water, the top layer of the sample was leveled with the edge of the mold, after the mold had been filled. To prevent excessive evaporation of the mixing water, and shrinkage strains brought on by the heat of the cement hydration process, the upper layer of the sample was covered with absorbent mats twenty-four hours after it was manufactured. A 12 h pre-treatment period was involved. The samples were then de-molded and kept in water, by EN 12390-2:2019-07 [27]. The laboratory conditions during the whole manufacturing process were, 21 °C temperature and 50% humidity.

2.4. Physical Properties Testing Methodology of Cement–Glass Composite Brick

To pinpoint the thermal characteristics of the hardened CGM, the thermal conductivity, thermal diffusivity, and specific heat of the hardened samples were studied. The ISOMET 2114 analyzer was used for all measurements (Applied Precision Ltd., Bratislava, Slovakia). The resistor heater on the analyzer's probe was close to the sample being tested, making it possible to measure how the material's temperature responded to heat flow pulses. Also employed was an analyzer, with a 60 mm diameter probe. Ten specimens, in the form of cubes, with dimensions of 150 mm × 150 mm × 150 mm, were evaluated, to determine all of the CGM's thermal characteristics. These samples were also used to test the density of the hardened samples. This was computed by dividing the mass by the sample volume. Also, the same procedure was performed on ten final CGCB specimens to compare the results, and indicate the influence of the interior PET-G scaffolding on the brick's properties.

2.5. Mechanical Properties Testing Methodology with a Digital Image Correlation (DIC)

The compression and bending strength tests for the CGCB and CGCB with polymeric scaffolding samples were performed using the Instron 8802 (Instron, Norwood, MA, USA) testing machine, shown in Figure 2. Bending tests were performed on a test rig, with a support distance of 120 mm. Both bending and compression tests were on the same samples (40 mm × 40 mm × 160 mm). All tests were conducted employing DIC measurements, utilizing a Dantec Dynamic (Dantec, Ulm, Germany) device and the ISTRA 4D software.

Figure 2. Instron 8802 servo-hydraulic pulsator for three-point bending and compression tests, equipped with the Dantec DIC system.

3. Results and Discussion

3.1. Physical Properties of Concrete–Glass Composite Brick

3.1.1. Density

The CGCB samples acquired a reduced density equivalent to 2051 kg/m^3, but the CGM sample had an average density of 2157 kg/m^3 [14]. Both concretes were assigned to the standard concrete class [28]. Using additional polymeric scaffolding caused a density reduction of about 5% (a difference of 106 kg/m^3). Such a phenomenon was caused by the significant mass difference between the cement–glass mortar and the PET-G material. Additionally, the tension between the plastic scaffolding and glass cullets indicated the presence of more air voids, that lowered the brick's density. Małek et al. [14] presented similar trends when using the three dimensional structure "3Dhon" for scaffolding. In their study, the final CGCB reached a density of 1982 kg/m^3, and was classified as a lightweight concrete, with D2.0 class [28]. This disparity results from the design of the interior scaffolding, which was thicker and took a greater mold volume than the gyroidal structure tested in this study. Despite this difference, it confirms a downward trend in terms of the density of the cement–glass mortar with the printed PET-G reinforcement. Also, the CGCB showed about the same density as traditional bricks, tested by Bautista-Marín et al. [29], without any additions (about 2200 kg/m^3).

3.1.2. Thermal Tests on CGM and CGCB

The hardened CGCB was tested for its thermal properties. The average values of thermal conductivity, thermal diffusivity, and specific heat were calculated based on ten measurements. As presented in Table 4, compared to the cement–glass mortar [14], the final CGCB showed about a 12% decrease in thermal conductivity (0.87 ± 0.05 W/mK), and about a 20% decrease in specific heat (1.31 ± 0.01 MJ/m^{3}K). Only thermal diffusivity increased after incorporating the PET-G scaffolding with the gyroidal structure into the matrix. The increase was from 0.61 ± 0.03 µm^2/s to 0.69 ± 0.03 µm^2/s. Based on available research [14], the gyroidal PET-G scaffolding showed similar trends to the 3Dhon scaffolding incorporated into a cement–glass mortar matrix. However, samples tested by Małek et al. [14], showed lower changes to the thermal properties: 9%, 10%, and 8% for thermal conductivity, specific heat, and thermal conductivity, respectively.

Table 4. Thermal properties of composite materials.

Sample Symbol	Thermal Conductivity [W/mK]	Thermal Diffusivity (µm^2/s)	Specific Heat (MJ/m^{3}K)
CGM [14]	0.99 ± 0.05	0.61 ± 0.03	1.64 ± 0.01
CGCB in this study	0.87 ± 0.05	0.69 ± 0.03	1.31 ± 0.01
CGCB by Małek et al. [14]	0.91 ± 0.05	0.66 ± 0.03	1.48 ± 0.01

3.2. Compressive and Bending Tests with the Use of DIC

3.2.1. Three-Point Bending Tests

The flexural strength was determined for samples oriented in the vertical and horizontal directions (shown in Figure 1). To better understand the influence of the scaffolding used, the results of the test (shown in Figure 3) include concrete samples without scaffolding (CGC vertical and horizontal), separated polymeric reinforcement (reinforcement vertical and horizontal), and the manufactured composite of concrete and polymeric scaffold (composite vertical and horizontal).

The measured flexural strength of CGC, oriented vertically and horizontally, reached values of 2.4 kN and 2.1 kN, respectively. It is seen that, in the vertical direction, samples are characterized by an almost two times greater beam deflection (from 0.22 mm to 0.4 mm). The use of the additional polymeric scaffolding inverted this phenomenon. It caused the properties to be more isotropic from the beam deflection point of view (the deviation of the representative curves is visibly smaller). Another important advantage of using polymeric reinforcement with a gyroidal geometry, is that it caused a 5% increase in the total force, in the case of both (horizontal and vertical) directions. Another significant statement could be made based on the DIC results (Figure 4). There is not any significant difference between the CGCB with or without polymeric scaffolding. This phenomenon allows us to state that, from a mechanical properties point of view, the material's behavior during static loading is similar to conventionally made bricks, without additional scaffolding. Such positive results could be justified because of the high level of the mortar's distribution between the polymeric structure cells. The glass particles went through the bigger cells in the designed scaffold, which positively affected the fracture behavior stability in both orientations for the tested samples. The visible crack course in Figure 4a can be compared to Figure 4c, and that in Figure 4b to Figure 4d, respectively. Such a phenomenon proves the stability of the material during static loading.

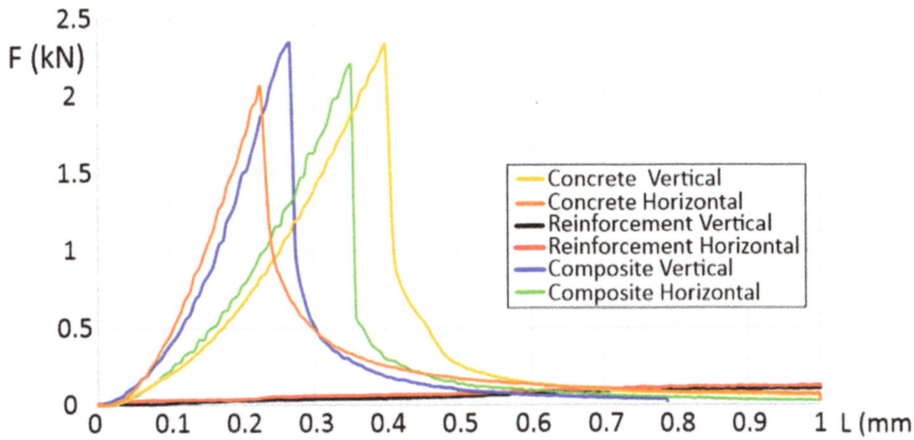

Figure 3. Force as a function of deformation in three-point bending testing.

CGCB showed greater than 5 MPa flexural strength, usually taken as a standard for masonry materials [30]. Its flexural strength of 5.90 MPa in the horizontal orientation, and 6.75 MPa in the vertical orientation, proves that it can be a replacement for traditional masonry materials as a novel eco-building material.

The obtained DIC results do not differ significantly from the data available in the literature. Shah et al. [31] analyzed mortar bricks using the DIC method during three-point bending tests. The authors registered a single crack tip along the sample part, with significant amounts of microcracks in the fracture zones that form ahead of the traction-free crack, a property of a quasi-brittle material. A visible increase in total deformation of horizontally oriented composite samples (Figure 3), decreases the typical behavior of cracking (which is proved by a more complex crack tip in the case of the horizontally oriented samples). Quite a different approach, more suitable for CGCB samples with AM scaffolding, was suggested by Wu et al. [32], where they used a cohesive crack concept, that allows a description of the strain-softening behavior of the material. This approach finds the relationship between the fracture process zone and the crack length. The results obtained utilizing the DIC method help to better study the behavior of the crack extension resistance curve of the material. Based on this method, and the DIC measurements, it could be observed that the length increased during crack propagation, but decreased after the fracture process zone fully developed at a crack extension.

3.2.2. Compressive Tests Results and Discussion

A positive effect of using internal polymeric scaffolding is also visible in the case of the compressive testing. The representative courses of each combination are shown in Figure 5. The reinforced CGCB is characterized by more isotropic properties than registered in the samples without any scaffolding. The courses of the curves for the horizontal CGCB samples, with and without scaffolding, are similar, and the same phenomenon is visible in the vertical direction. In the case of the vertical direction, the courses of the curves are almost the same. This shows that this type of internal scaffolding has much better properties than those registered in the case of our own previous research [14].

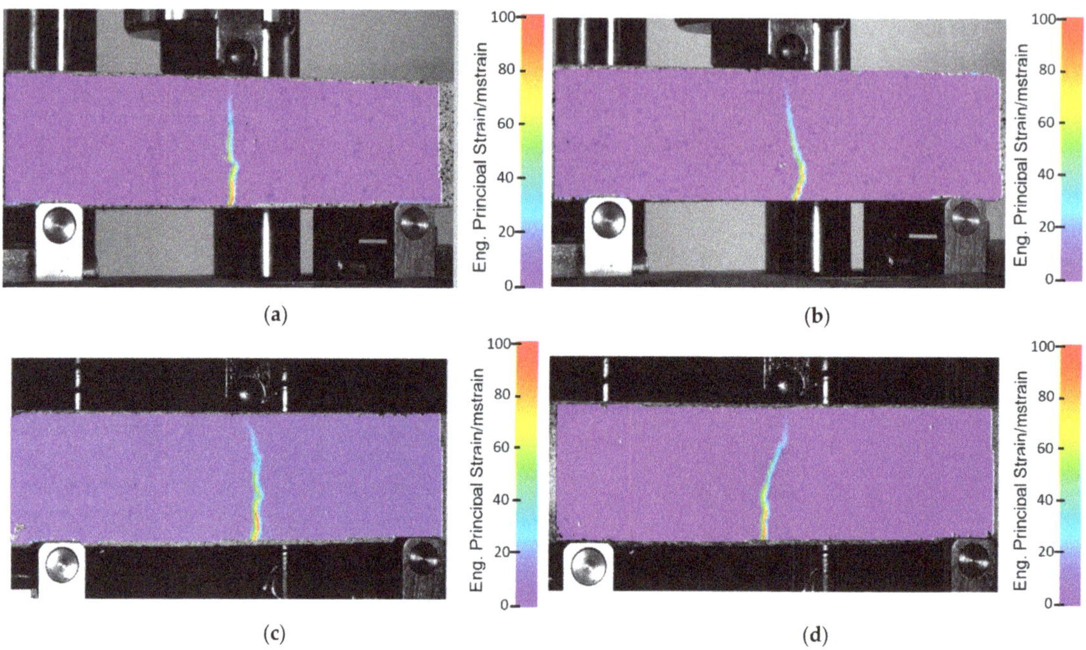

Figure 4. Results of the DIC analyses during flexural strength tests of CGCB samples in (**a**) horizontal, and (**b**) vertical orientation; and CGCB with scaffolding in (**c**) horizontal, and (**d**) vertical orientation.

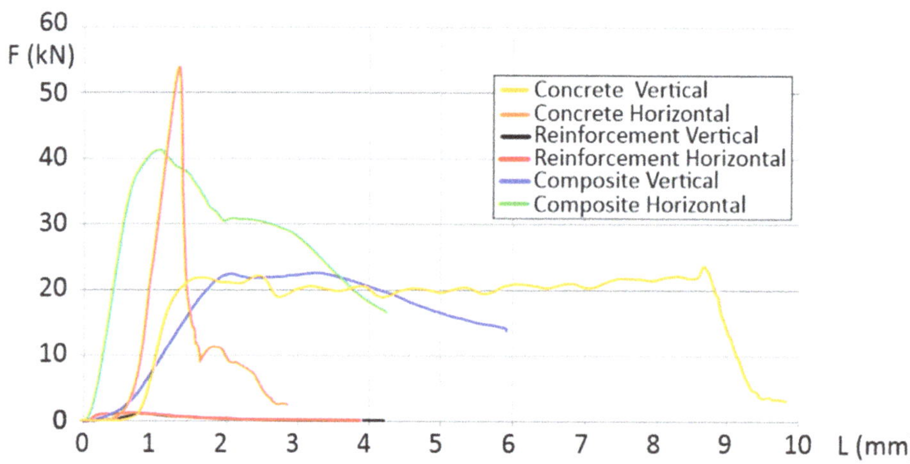

Figure 5. Force as a function of deformation in compression testing.

Additionally, the designed cement–glass composite brick's compressive strength is in line with traditional bricks, tested by Zuo et al. [33]. They showed about 24 MPa compressive strength, compared to 25.6 MPa for CGCB in the horizontal orientation. The value reported for vertical orientation was much lower—14.4 MPa—because of the asymmetry of the sample. The glass grain sizes varied between horizontal and vertical orientations due to their random distribution.

The main difference was related to the polymeric structure's internal cell dimensions, and the gyroidal structure used in this research allowed for a significantly better glass particle distribution. Such an approach obtains the essential advantage of composite structures—a combination of the properties characterizing both materials. The described phenomenon is depicted in Figure 6, in the microscopic fracture images. Each cell has been evenly filled with glass particles.

(**a**) (**b**)

Figure 6. The fracture surface of the tested samples: (**a**) without scaffolding (**b**) with internal, polymeric scaffolding.

Another analysis connected with the DIC measurements, revealed a positive influence of the scaffolding structure used, partially shown in the compressive testing curves. In the case of the reinforced samples, the increased strain areas are significantly larger than in samples without additional scaffolding. This kind of phenomenon is visible in both the vertical (Figure 7i) and horizontal (Figure 7l) orientations. This answers why additional scaffolding changed the material's characteristics to being more isotropic.

A practical method, that could be used for a proper description of this kind of material's behavior, was described by Zhou et al. [34], for rock-like specimens. The stress–strain curves in this methodology were divided into two stages:

- the strain-softening stage;
- the residual strength stage.

The strain-softening stage is characterized by rapid crack evolution and further propagation, leading to a rapid drop in the axial stress. With the increase in uniaxial compressive loads, the axial stresses decrease to the residual strengths in the materials with different brittleness indices (i.e., polymeric scaffold or waste glass reinforcement). Such an approach would be possible by analyzing particular cells of the scaffold.

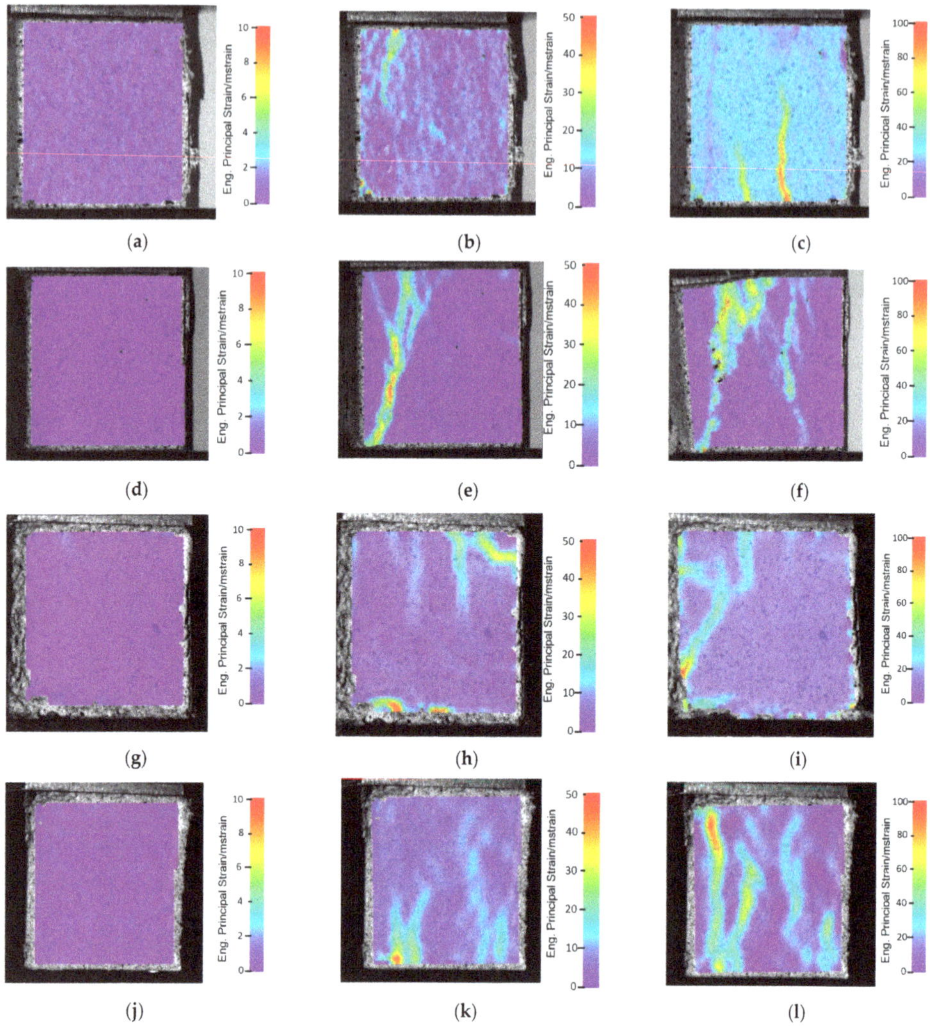

Figure 7. DIC images of compressive strength tests of samples without scaffolding (**a**–**f**), and with scaffolding (**g**–**l**): in vertical arrangement (**a**) at the start, (**b**) after 1 mm of strain, (**c**) before sample breaking; and in horizontal arrangement (**d**) at the start, (**e**) after 1 mm of strain, (**f**) before sample breaking; and samples with scaffolding in vertical arrangement (**g**) at the start, (**h**) after 1 mm of strain, (**i**) before sample breaking; and in horizontal arrangement (**j**) at the start, (**k**) after 1 mm of strain, (**l**) before sample breaking.

4. Conclusions

The main aim of this research was to determine the mechanical and thermal behaviors of the proposed CGCB bricks. An additional purpose was to compare scaffolded and non-scaffolded test samples. Such an approach allowed for a deep analysis of the use of additional polymeric structures. The results revealed the positive physical properties (thermal and mechanical) of a new type of concrete with glass particles, additionally reinforced by an additively manufactured polymeric structure. The analyses conducted highlight the positive influence of using waste materials, such as glass particles or recycled PET-G. The combination of these two materials in building engineering solutions is also possible. Based on the mixtures law, it could be additionally concluded that the addition of such kinds of waste materials increases the chemical resistance of the produced bricks. The results of this research allow us to draw the following conclusions:

1. The cement–glass composite brick, with a printed PET-G gyroidal structure, has better thermal properties than the cement–glass mortar itself, as thermal conductivity after the modification decreased by 12%, and specific heat was reduced by 20%. The thermal diffusivity, on the other hand, increased by 13%.
2. After incorporating the gyroidal structure from PET-G into the cement–glass mortar matrix, the density changed from 2157 kg/m^3 to 2051 kg/m^3.
3. The use of the AM gyroidal structure made the concrete brick's mechanical properties more isotropic, with up to 50% lower beam deflection during the three-point bending test.

To obtain more stable properties of the material, in different directions, it is necessary to use polymeric structures with proper cell dimensions—it is essential to allow the easy flow of the glass particles in the material's volume.

For further research, it is necessary to create a structure that could improve the mechanical properties of the produced bricks. Such a positive effect in this research (as more isotropic properties), could be improved by increasing the volume of the scaffolding material.

The novel cement-glass composite brick, made from approximately 86% of waste, is an example of the reuse of rubbish from other industries in the construction sector. It should be noted that the durability of the final CGCB is comparable to traditional bricks, while ensuring increased thermal insulation at the same time.

Author Contributions: Conceptualization, M.M. and J.K.; methodology, J.Ł., J.T., I.S. and W.Ł.; software, I.S. and M.J.; validation, M.M., K.G. and M.J.; formal analysis, J.K.; investigation, J.Ł., W.Ł., J.T., I.S. and J.K.; resources, J.K.; data curation, M.M., I.S. and J.Ł.; writing—original draft preparation, M.J., J.K. and W.Ł.; writing—review and editing, J.K. and M.M.; visualization, I.S. and J.Ł.; supervision, M.M. and J.K.; project administration, M.M.; funding acquisition, J.K. All authors have read and agreed to the published version of the manuscript.

Funding: This work and APC was financed by Military University of Technology under research project UGB 22 830. Additionally, a part of the research was funded by the Dean of the Faculty of Civil Engineering and Geodesy of the Military University of Technology as part of grant no. UGB/813/2023.

Institutional Review Board Statement: Not applicable.

Informed Consent Statement: Not applicable.

Data Availability Statement: Data available on request.

Conflicts of Interest: The authors declare no conflict of interest.

References

1. Kamiński, K.; Plis, A. Research on Compresive Strength and Strain of Expansive Concrete. *Appl. Mech. Mater.* **2015**, *797*, 79–87.
2. Available online: https://tradingeconomics.com/commodity/steel (accessed on 20 February 2022).
3. Salazar, B.; Aghdasi, P.; Williams, I.D.; Ostertag, C.P.; Taylor, H.K. Polymer lattice-reinforcement for enhancing ductility of concrete. *Mater. Des.* **2020**, *196*, 109184. [CrossRef]
4. Morretton, E.; Vignat, F.; Pourroy, F.; Marin, P. Impacts of the settings in a design for additive manufacturing process based on topological optimization. *Int. J. Interact. Des. Manuf.* **2019**, *13*, 295–308. [CrossRef]
5. Doutre, P.-T.; Morretton, E.; Vo, T.H.; Marin, P.; Pourroy, F.; Prudhomme, G.; Vignat, F. Comparison of some approaches to define a CAD model from topological optimization in design for additive manufacturing. *Lect. Notes Mech. Eng.* **2017**, *1*, 233–240. [CrossRef]
6. Gardan, N.; Schneider, A.; Gardan, J. Material and process characterization for coupling topological optimization to additive manufacturing. *Comput. Aided. Des. Appl.* **2015**, *13*, 39–49. [CrossRef]
7. Xu, B.; Han, Y.; Zhao, L.; Xie, Y.M. Topological optimization of continuum structures for additive manufacturing considering thin feature and support structure constraints. *Eng. Optim.* **2020**, *53*, 2122–2143. [CrossRef]
8. Shifeng, W.; Shuai, L.; Qingsong, W.; Yan, C.; Sheng, Z.; Yusheng, S. Effect of molten pool boundaries on the mechanical properties of selective laser melting parts. *J. Mater. Process. Technol.* **2014**, *214*, 2660–2667. [CrossRef]

9. Xu, Y.; Zhang, H.; Gan, Y.; Šavija, B. Cementitious composites reinforced with 3D printed functionally graded polymeric lattice structures: Experiments and modelling. *Addit. Manuf.* **2021**, *39*, 101887. [CrossRef]
10. Qin, S.; Cao, S.; Yilmaz, E.; Li, J. Influence of types and shapes of 3D printed polymeric lattice on ductility performance of cementitious backfill composites. *Constr. Build. Mater.* **2021**, *307*, 124973. [CrossRef]
11. Fadeel, A.; Mian, A.; Al Rifaie, M.; Srinivasan, R. Effect of Vertical Strut Arrangements on Compression Characteristics of 3D Printed Polymer Lattice Structures: Experimental and Computational Study. *J. Mater. Eng. Perform.* **2018**, *28*, 709–771. [CrossRef]
12. Salazar, B.; Williams, I.; Aghdasi, P.; Ostertag, C.; Taylor, H. Bending and Crack Characteristics of Polymer Lattice-Reinforced Mortar. In Proceedings of the International Congress on Polymers in Concrete (ICPIC 2018), Washington, DC, USA, 29 April –1 May 2018; pp. 261–266. [CrossRef]
13. Lin, A.; Tan, Y.K.; Wang, C.-H.; Kua, H.W.; Taylor, H. Utilization of waste materials in a novel mortar–polymer laminar composite to be applied in construction 3D-printing. *Compos. Struct.* **2020**, *253*, 112764. [CrossRef]
14. Małek, M.; Grzelak, K.; Łasica, W.; Jackowski, M.; Kluczyński, J.; Szachogłuchowicz, I.; Torzewski, J.; Łuszczek, J. Cement-glass composite bricks (CGCB) with interior 3D printed PET-G scaffolding. *J. Build. Eng.* **2022**, *52*, 104429. [CrossRef]
15. Grzelak, K.; Łaszcz, J.; Polkowski, J.; Mastalski, P.; Kluczyński, J.; Łuszczek, J.; Torzewski, J.; Szachogłuchowicz, I.; Szymaniuk, R. Additive Manufacturing of Plastics Used for Protection against COVID19—The Influence of Chemical Disinfection by Alcohol on the Properties of ABS and PETG Polymers. *Materials* **2021**, *14*, 4823. [CrossRef]
16. Woern, A.L.; Byard, D.J.; Oakley, R.B.; Fiedler, M.J.; Snabes, S.L.; Pearce, J.M. Fused Particle Fabrication 3-D Printing: Recycled Materials' Optimization and Mechanical Properties. *Materials* **2018**, *11*, 1413. [CrossRef] [PubMed]
17. Gu, H.; AlFayez, F.; Ahmed, T.; Bashir, Z. Poly(ethylene terephthalate) Powder—A Versatile Material for Additive Manufacturing. *Polymers* **2019**, *11*, 2041. [CrossRef]
18. Dolzyk, G.; Jung, S. Tensile and Fatigue Analysis of 3D-Printed Polyethylene Terephthalate Glycol. *J. Fail. Anal. Prev.* **2019**, *19*, 511–518. [CrossRef]
19. Latko-Durałek, P.; Dydek, K.; Boczkowska, A. Thermal, Rheological and Mechanical Properties of PETG/rPETG Blends. *J. Polym. Environ.* **2019**, *27*, 2600–2606. [CrossRef]
20. *EN 197-1:2012*; Cement—Part 1: Composition, Specifications and Conformity Criteria for Common Cements. European Committee for Standardization: Brussels, Belgium, 2012.
21. *EN 196-6:2019-01*; Methods of Testing Cement—Part 6: Determination of Fineness. European Committee for Standardization: Brussels, Belgium, 2019.
22. *EN 196-1:2016-07*; Methods of Testing Cement—Part 1: Determination of Strength. European Committee for Standardization: Brussels, Belgium, 2016.
23. Rudnicki, T. The method of aggregate skeleton in self compacting concrete designing with segment regression. *CWB* **2016**, *1*, 10–19. [CrossRef]
24. *EN 12620+A1:2010*; Aggregates for Concrete. European Committee for Standardization: Brussels, Belgium, 2010.
25. Górażdże Group: Cement, Concrete, Aggregate. Technical Data Sheet CEM I 42.5 R NA. Available online: http://www.gorazdze.pl (accessed on 15 June 2022).
26. Rudnicki, T. Functional Method of Designing Self-Compacting Concrete. *Materials* **2021**, *14*, 267. [CrossRef] [PubMed]
27. *EN 12390-2:2019-07*; Testing Hardened Concrete—Part 2: Making and Curing Specimens for Strength Tests. European Committee for Standardization: Brussels, Belgium, 2019.
28. *EN 12390-7:2019-08*; Testing Hardened Concrete—Part 7: Density of Hardened Concrete. European Committee for Standardization: Brussels, Belgium, 2008.
29. Bautista-Marín, J.; Esguerra-Arce, A. Use of an industrial solid waste as a pigment in clay bricks and its effects on the mechanical properties. *Constr. Build. Mater.* **2021**, *306*, 124848. [CrossRef]
30. De Rosa, B.; Cultrone, G. Assessment of two clayey materials from northwest Sardinia (Alghero district, Italy) with a view to their extraction and use in traditional brick production. *Appl. Clay Sci.* **2014**, *88–89*, 100–110. [CrossRef]
31. Shah, S.G.; Kishen, J.M.C. Fracture Properties of Concrete-Concrete Interfaces Using Digital Image Correlation. *Exp. Mech.* **2011**, *51*, 303–313. [CrossRef]
32. Wu, Z.M.; Rong, H.; Zheng, J.J.; Xu, F.; Dong, W. An Experimental Investigation on the FPZ Properties in Concrete Using Digital Image Correlation Technique. *Eng. Fract. Mech.* **2011**, *78*, 2978–2990. [CrossRef]
33. Zuo, H.; Wang, C.; Zhang, J.; Shao, J.; Zhao, Y.; Jiao, K. Comparison of oxidation behaviors of novel carbon composite brick with traditional carbon brick. *Ceram. Int.* **2015**, *41*, 7929–7936. [CrossRef]
34. Zhou, X.P.; Wang, Y.T.; Zhang, J.Z.; Liu, F.N. Fracturing Behavior Study of Three-Flawed Specimens by Uniaxial Compression and 3D Digital Image Correlation: Sensitivity to Brittleness. *Rock Mech. Rock Eng.* **2019**, *52*, 691–718. [CrossRef]

materials MDPI

Article

Annual Plants and Thermoplastics in the Production of Polymer and Lignocellulose Boards

Aleksandra Banaszak [1,*], Magdalena Woźniak [2], Dorota Dziurka [1] and Radosław Mirski [1]

[1] Department of Mechanical Wood Technology, Poznan University of Life Sciences, Wojska Polskiego 28, 60-627 Poznań, Poland; dorota.dziurka@up.poznan.pl (D.D.); radoslaw.mirski@up.poznan.pl (R.M.)
[2] Department of Chemistry, Faculty of Forestry and Wood Technology, Poznan University of Life Sciences, Wojska Polskiego 75, 60-625 Poznań, Poland; magdalena.wozniak@up.poznan.pl
* Correspondence: aleksandrabanaszak@op.pl

Abstract: This study investigated the mechanical, physical, and thermal properties of three-layer particleboards produced from annual plant straws and three polymers: polypropylene (PP), high-density polyethylene (HDPE), and polylactic acid (PLA). The rape straw (*Brassica napus* L. var. Napus) was used as an internal layer, while rye (*Secale* L.) or triticale (*Triticosecale* Witt.) was applied as an external layer in the obtained particleboards. The boards were tested for their density, thickness swelling, static bending strength, modulus of elasticity, and thermal degradation characteristics. Moreover, the changes in the structure of composites were determined by infrared spectroscopy. Among the straw-based boards with the addition of tested polymers, satisfactory properties were obtained mainly using HDPE. In turn, the straw-based composites with PP were characterized by moderate properties, while PLA-containing boards did not show clearly favorable properties either in terms of the mechanical or physical features. The properties of straw–polymer boards produced based on triticale straw were slightly better than those of the rye-based boards, probably due to the geometry of the strands, which was more favorable for triticale straw. The obtained results indicated that annual plant fibers, mainly triticale, can be used as wood substitutes for the production of biocomposites. Moreover, the addition of polymers allows for the use of the obtained boards in conditions of increased humidity.

Keywords: polymer–straw boards; thermoplastic polymers; annual plants; mechanical properties

Citation: Banaszak, A.; Woźniak, M.; Dziurka, D.; Mirski, R. Annual Plants and Thermoplastics in the Production of Polymer and Lignocellulose Boards. *Materials* **2023**, *16*, 4400. https://doi.org/10.3390/ma16124400

Academic Editors: Marcin Małek and Janusz Kluczyński

Received: 10 May 2023
Revised: 12 June 2023
Accepted: 13 June 2023
Published: 15 June 2023

1. Introduction

Replacing the traditionally used particles in particleboard production with particles of annual crops is still a current topic. The reasons that wood particles are being substituted are, among others, the high prices, storage problems, intense market competition, and lack of suitable quality materials [1]. In turn, annual products are available in abundant volume throughout the world. It is worth highlighting that the forecast for annual surplus straw in Poland in 2025 is circa 14 thousand tons [2]. The most widely available straws in Poland are cereal straws (73.5%) and rape straws (9.1%) [3].

The literature describes the use of various lignocellulosic materials as a substitute for wood in particleboard manufacturing. The particleboards were made of wheat straw, rape straw, rice straw, corn straw, reed stalk chips, rye straw, triticale straw or ground-nut shell, and rice husk wastes [4–16]. The data in the literature also showed that white mustard can be applied as a good alternative to wood in particleboard production [17]. Moreover, evening primrose straw, kiwi, grape, and tomato stalks and coffee or tea waste are interesting options in the replacement of wood chips in the manufacturing of particleboards [18–21]. Although many studies have been conducted on the use of various lignocellulosic materials for the production of oriented strand boards (OSBs), not all of the obtained materials provide the physical and mechanical properties required in construction (type five according to the EN 310 [22]). In our previous research, we analyzed the possibility of replacing pine chips with chips from annual plants (rye, triticale, rape, reed, and corn straw) in the outer layers of boards intended for the construction and furniture industry [11]. The use of annual plant straw was particularly favorable for the modulus of elasticity and smaller thickness swelling as compared with the pine boards. Moreover, the results indicated that, in specific conditions, all tested plants could serve as a partial substitute for wood chips in the external layers of particleboards [11]. Annual plants are an attractive alternative to wood chips in the production of chipboards due to, among other things, their low price, lower hygroscopicity, and specific gravity or better thermal and acoustic isolation [9,23–25]. On the other hand, straw has a larger amount of wax on the surface, which causes difficulties in using conventional adhesives. However, the modification of straws or adhesives such as pMDI (polymeric diphenylmethane diisocyanate) is the solution to this problem [6,26,27].

The particles of straw were also successfully used as a filler in a WPC (wood–plastic composite), using, among other straws, rape, sisal, hemp, jute fibers, and rice straw [28–30]. Mainly WPC is manufactured by methods used in the plastic industry. The exciting approach is flat pressing, which was used by numerous researchers [31–37]. The OSB or particleboard with polymer addition in inner layers was characterized by better physical properties, which maintained good mechanical properties. The plastic industry is very powerful, and the annual waste generated after using plastic products is still a big concern in our society. Using surplus straws and plastic polymers in particleboard is one idea to face the market's needs and gain a fully valuable product.

In our previous research, we investigated the effect of various types of straw (rye, triticale, and rape) and various type of thermoplastics (HDPE, low-density polyethylene (LDPE), polystyrene (PS), and PP) on the properties of five-layer boards [31]. The results showed that the properties of the obtained boards strongly depended on both kinds of the polymer and straw used [31]. The research described by Mihai et al. [38,39] indicated that triticale straw has great potential in the production of thermoplastic composites. Moreover, the triticale content, the presence of maleic anhydride grafted polypropylene as a coupling agent, and the addition of calcium oxide as a reactive additive have an influence on the properties of PP/triticale straw biocomposites [38,39]. Moreover, biocomposites based on triticale straw and PLA were characterized by good properties, and due to their biodegradability of both PLA and triticale, they are more environmentally friendly than synthetics based on thermoplastics [40]. Moreover, the PP and rye-husk-based composites showed good performance compared to softwood composites, including better elongation at break and better Charpy impact strength [41].

As shown above, the application of straw from various annual plants, including triticale and rye, as a substitute for wood in the production of composites has been reported in the literature [30–37]. In this paper, the characterization of particleboards made from straw (triticale and rye) with the addition of three polymers, namely PP, HDPE, and PLA, is reported. To the best of the authors' knowledge, this is the first report about the mechanical, physical, and thermal properties of three-layer polymer–straw boards produced based on triticale and rye and three different polymer matrices and with rape straw as an inner layer, with different moisture contents for the layers.

2. Materials and Methods

2.1. Materials

Three-layer boards were manufactured from particles of rape straw (*Brassica napus* L. var. Napus) as an internal layer and particles of cereal straw, i.e., triticale (*Triticosecale* Witt.) and rye (*Secale* L.), with a 30% addition of polymers as an external layer. The control boards were made without the addition of polymers. Polymers used as an addition in external layers were polypropylene (PP), high-density polyethylene (HDPE), and polylactic acid (PLA). PP and HDPE were obtained from bags for lunch and garbage, respectively. The bags are widely available at the most significant discounts in Poland. The bags were cut into thin strips 5 mm wide and 20–30 mm long. PLA was used in the form of granules.

In Table 1, the structure of the boards is presented. The rye and the triticale straws were prepared traditionally, as described in detail in our previous work [11]. The very fine fraction passing through a sieve with a 0.5 × 0.5 mm^2 mesh was removed. The straw particles prepared this way had a bulk density of 60 ± 2.5 kg/m^3. A significant difference for the chosen straws was the width of the particles; i.e., wider and slightly thicker particles are characteristic of triticale [33]. To increase the compression of the middle layer, the rapeseed chopped strands were sieved through the sieve with a mesh of 2 × 2 mm^2. This decreased the bulk density from about 80 kg/m^3 to 48 ± 3.2 kg/m^3. That was important because of the higher humidity of this layer. Due to adding polymers, the outer layers lost their initial low bulk density and vulnerability to compression during pressing.

Table 1. Structure of polymer-enriched boards.

Symbol	External Layers	Core Layer	Content of Polymer	Polymer
Control 1 (reference sample)	Rye straw	Rape straw	-	-
RP	Rye straw	Rape straw	30%	PP
RE	Rye straw	Rape straw	30%	HDPE
RA	Rye straw	Rape straw	30%	PLA
Control 2 (reference sample)	Triticale straw	Rape straw	-	-
TP	Triticale straw	Rape straw	30%	PP
TE	Triticale straw	Rape straw	30%	HDPE
TA	Triticale straw	Rape straw	30%	PLA

2.2. *Board Manufacturing*

The chips were glued with the same amount (5% of dry weight) of pMDI (polymeric diphenylmethane diisocyanate, Ongronat® 2100, BorsodChem Group, Kazincbarcika, Hungary) in all layers and all types of boards. Ongronat® 2100 is a brown (20 °C, 1013 hPa) non-flammable liquid of a density of 3 g/cm^3, viscosity of 210 mPa·s, contractual dry matter content of 100%, NCO content of 30.6%, and chlorine hydrolytic content of 127 ppm. Straws of annual plants show low gluability for commonly used formaldehyde-based adhesives. In addition, adhesion promoters are indicated for bonding lignocellulosic particles and polymers [6,8,14,26,27]. Therefore, based on the literature data and our previous studies, we decided to apply pMDI in the production of boards [11,36,37].

The material intended for the external layers was dried up to 2% humidity, and that for the core layer was up to 9%. The material was dried in a laboratory drum dryer (DAN LAB, Białystok, Poland). The moisture distribution is not typical for particleboards because the outer layers had much lower humidity than the middle layer. However, good results of combining thermoplastic polymers with lignocellulosic particles are obtained without water content in the plant material. On the other hand, pMDI itself requires higher humidity than the assumed 2%. It was assumed, however, that a lower straw moisture content than recommended in industrial practice for pMDI crosslinking can be compensated for by better anchoring the polymers [9,11].

The proportion of external and core layers was 3:7. The pressing conditions are presented in Table 2.

Table 2. Pressing conditions.

Parameter	Value
The temperature of hotplates	205 °C
Maximum pressure	2.0 N/mm^2
Pressing time	40 s/mm of board thickness
Board thickness	15 mm
Density	575 kg/m^3

The outer layers needed to be overheated to allow for the glue's crosslinking inside the board. The pressing time was more than doubled.

2.3. Board Testing

After the boards were conditioned for seven days at 21 ± 1 °C and relative air humidity of 55 ± 5%, the produced boards were tested in terms of the following parameters, according to the relevant standards:

- Bending strength (MOR) and modulus of elasticity (MOE) according to EN 310 [22];
- Internal bond (IB) according to EN 319 [42];
- Internal bond after the boiling test (V-100) according to EN-1087-1 [43];
- Thickness swelling (TS) after 24 h according to EN 317 and water absorption (WA) [44].

The investigations of water resistance and mechanical properties involved 10-to-16 samples in each variant. The number of samples was in accordance with the relevant standards indicated above.

The quality of the melted polymers and their interaction with straw particles were also analyzed using computed tomography. For this purpose, samples from boards made from triticale straw were scanned. Scanning was performed with a Hyperion X9Pro CT scanner with a resolution of 0.15 mm, at a tube voltage of 90 kV, a point resolution of 68 m, and a maximum imaging field of 13 cm × 16 cm (MyRay, Via Bicocca, Imola, Bologna, Italy). The Hounsfield (HU) scale was used to estimate the particleboard structure. The Hounsfield unit is a relative quantitative measurement of radio frequency density used by radiologists to interpret computed tomography (CT) images. Although this is not a very accurate measure, it allows for determining changes in the structure of the particleboard.

2.4. Attenuated Total Reflection–Fourier-Transform Infrared Spectroscopy (ATR-FTIR)

The spectra of tested boards were recorded by a Nicolet iS5 spectrophotometer (Thermo Fisher Scientific, Waltham, MA, USA) with Fourier transform, a deuterated triglycine sulfate (DTGS) detector, and an attenuated total reflection (ATR) attachment. The spectra were recorded over the range of 4000–500 cm^{-1}, at a resolution of 4 cm^{-1} and with 32 co-added scans. Ten measurements for each sample were recorded by re-sampling at different locations across the entire sample.

2.5. Thermogravimetric Analysis (TGA)

The thermogravimetric analysis (TGA) of boards was evaluated using a Netzsch STA 449 F5 Jupiter apparatus (Erich NETZSCH GmbH and Co. Holding KG, Selb, Germany). The board samples of 20 ± 1 mg were heated at the rate of 5 °C/min to the assumed final temperature of 600 °C in the atmosphere of helium flowing through the furnace space at a rate of 15 mL/min. Thermogravimetric (TG) curves and differential thermogravimetric (DTG) curves were recorded on the thermograms.

2.6. Statistical Analysis

The results were analyzed using STATISTICA 13.0 package (StatSoft Inc., Tulsa, OK, USA). The performed analysis was based on ANOVA (analysis of variance), and homogeneous groups were distinguished with the use of Tukey's HSD (honestly significant difference) test. Homogeneous groups are marked with lowercase letters. The results were analyzed at a significance level of $p = 0.05$.

3. Results and Discussion

The critical feature related to the mechanical properties of wood-based materials is a strong relation with density. In the manufacturing process, it was assumed that the density of the dry boards would be 575 kg/m^3 at a moisture content of 6.2 (SD = 0.39). Their density should be about 610 kg/m^3 during mechanical tests. As seen in Figure 1, the data show that the boards had a density close to the established one; however, according to the ANOVA statistical analysis, the densities of individual variants differed. The difference between the variants did not exceed 30 kg/m^3. The similar density of individual samples and, thus, the low SD (standard deviation) values, indicated statistically significant differences. The so-called conservative LSD (Least Significant Difference) test confirmed this, the results of which are also shown in Figure 1. The analysis using a less conservative post hoc test, which is more common in this type of analysis, i.e., Tuckey's HSD, did not show statistically significant differences in the density of individual types of boards. This is important because it is easier to analyze the behavior of the boards during the tests of their mechanical properties, as well as swelling or water absorption.

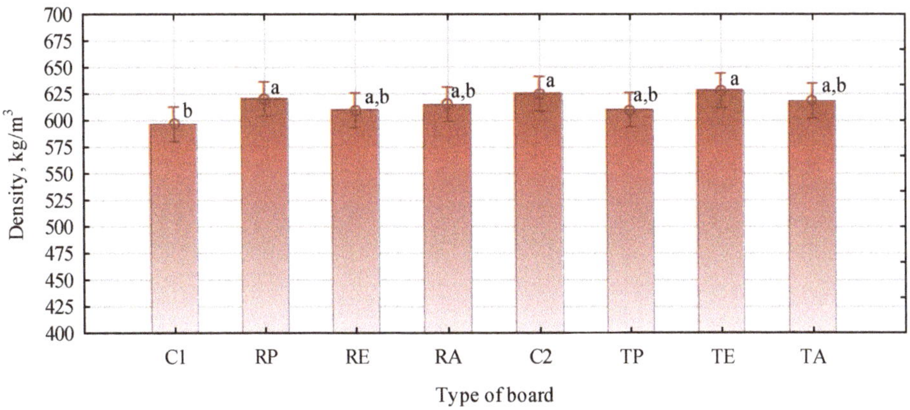

Figure 1. Densities of the produced boards: C1, control with rye straw; C2, control with triticale straw; R, rye straw; T, triticale straw; P, PP; E, HDPE; A, PLA. Values denoted with identical letters do not differ significantly in LSD test, $F(7, 88) = 1.4982$, $p = 0.017834$.

As expected, based on previous studies [11] the mechanical properties of boards made of triticale straw particles were slightly higher than boards made of rye straw particles. In addition, good mechanical properties were observed for single-layer boards made from rape particles [45]. This was applied primarily to the bending strength and modulus of elasticity. In the case of the static bending strength, a variance ANOVA was analyzed. It showed that boards made of triticale straw were characterized by a higher static bending strength than boards made of rye straw particles (Figure 2a). However, from a technological point of view, these differences were insignificant, as they did not exceed 4%. The type of polymer used affects the static bending strength of the manufactured boards more (Figure 2b). In this case, we observed an apparent increase in the strength of boards made with HDPE and a noticeable decrease in the strength of boards made with PLA. These changes were at +15% in the case of boards with HDPE, and −10% with PLA.

The statistical analysis of the interaction between the type of straw and the polymer shows no such relation (Figure 3). There were no statistically confirmed premises to recognize that the static bending strength of straw boards with the addition of polymers is related to the type of straw. Therefore, it can be assumed that some polymers work better with a given straw and others work worse. The data presented in Figure 3 indicate that only boards made of triticale with the addition of PP showed a lower bending strength than those based on rye straw. Interestingly, these boards had a very low modulus of elasticity (Figure 4). Their modulus of elasticity was more than 2500 N/mm^2, which can be considered high. The boards of wood chips of a similar thickness and density had a modulus of elasticity of 2200 N/mm^2. However, compared to the C2 control board, the modulus of elasticity of the TP-type board was nearly 15% lower. The changes in modulus of elasticity boards with polymers were also observed in other studies [36,37,46,47].

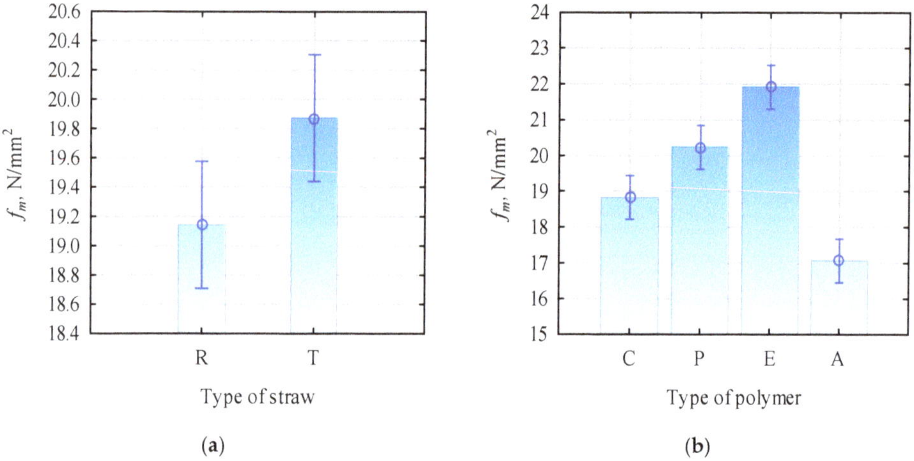

Figure 2. (**a**) Two-way-ANOVA effect of straw type, $F(1, 88) = 5.6101$, $p = 0.02005$. (Whiskers indicate 0.95 confidence intervals.) (**b**) Two-way-ANOVA effect of polymer type, $F(3, 88)=44.850$, $p = 0.0000$. (Whiskers indicate 0.95 confidence intervals.) C, control; R, rye straw; T, triticale straw; P, PP; E, HDPE; A, PLA.

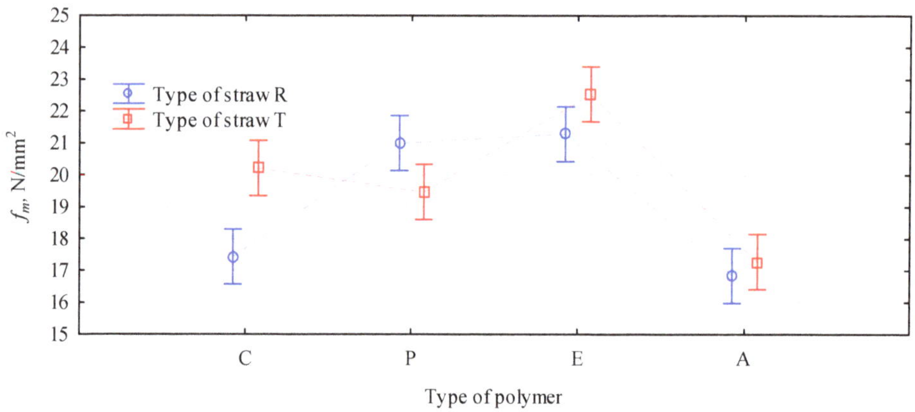

Figure 3. Graph of the interaction of the influence of the type of polymer and the type of straw on the bending strength, $F(3, 88) = 8.4941$, $p = 0.00005$. C, control; P, PP; E, HDPE; A, PLA.

The modulus of elasticity of the boards produced in the proposed manner was relatively untypical and deviated from expectations. The board with polymers of a relatively high strength showed a lower modulus of elasticity than those with lower strength. The exception was TE-type boards, i.e., with external layers made of triticale straw and HDPE. In their case, a high bending strength was correlated with high stiffness. There may be a reduction in the stiffness of straw boards produced with thermoplastic polymers since they naturally have lower stiffness. However, boards based on straw particles or wood, made with about 20% polymers, do not have to behave like typical polymer composites. In the case of the analyzed three-layer boards, the quality of the outer layers determines the stiffness. It depends on the quality of the polymer–straw bond, the distribution of polymers in the layer structure, the degree of compaction, and its structure. It should be recognized that the stiffness of straw–polymer boards determined by the modulus of elasticity is most favorably influenced by HDPE, and it is the least favorably influenced by PP. Both polymers originated from packaging crushing and had a similar form after grinding. Straw particles, unlike wood, do not have a good structure for anchoring thermoplastic polymers. In the case of boards made of pine chips and thermoplastic polymers, Borysiuk [34] obtained good results. However, the specialized treatments suitable for wood were different from straw. The smooth surface of straws from the grass genus and the spongy straw from the cabbage family do not create places for the mechanical embedding of molten polymers. In their case, the agents acting as adhesion promoters, the appropriate degree of fragmentation of both materials, and the technological parameters of producing such boards play an essential role.

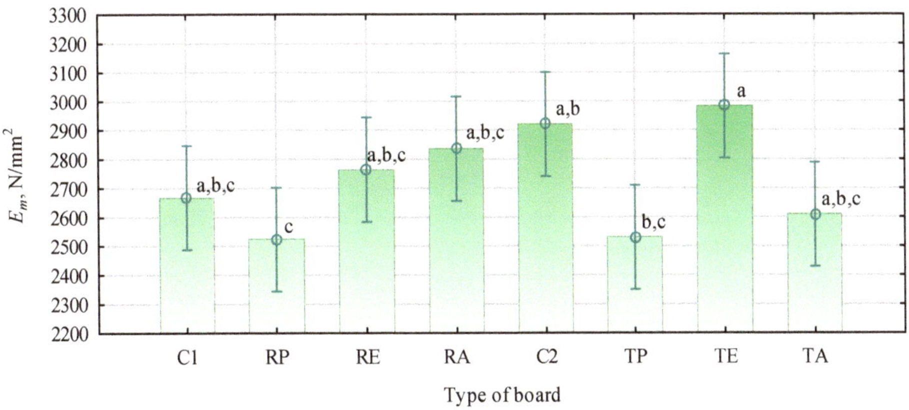

Figure 4. Modulus of elasticity of the produced boards: C1, control with rye straw; C2, control with triticale straw; R, rye straw; T, triticale straw; P, PP; E, HDPE; A, PLA. Values denoted with identical letters do not differ significantly in HSD Tukey's test, $F(7, 88) = 3.7348$, $p = 0.00138$.

The fulfillment of the first two factors may be evidenced by the swelling of the boards to the thickness after soaking in water. If the lignocellulosic material and thermoplastic polymer are properly and permanently merged, then the swelling of the boards produced in this way should be lower. The data presented in Figure 5 show that all boards with the addition of thermoplastic polymers to the outer layers had lower swelling than the control boards. The swelling of the control boards was about 31%, while that of the straw–polymer boards was about 21%. Generally, a positive effect is obtained when the swelling of the straw–polymer boards is lower than 24%. This condition still needed to be met by boards with PLA, and it only looked slightly better in the case of boards with PP and triticale straws. On the other hand, in the case of boards with HDPE, the swelling of boards, regardless of the type of straw, slightly exceeded 15%, which is about 50% lower than that of the control boards.

The swelling determined after 24 h of soaking in water is correlated with water absorption (Figure 6). Of course, the more water the board absorbs, the more intensively it swells; however, to absorb water intensively, it must have favorable conditions. On the one hand, there is high porosity, and on the other, there is easy access to the lignocellulosic structure. Thus, assuming that the boards were of a similar density, the possibility of penetration was similar. In contrast, with better coverage of the chopped straw with polymers and better compactness due to board gluing, its susceptibility to water absorption inside the lignocellulosic material was lower. The higher resistance to water absorption was also observed for the wood–plastic composite that was flat-pressed [48]. Generally, it can be assumed that the addition of polymers improves the water resistance of particleboard regardless of lignocellulose particles.

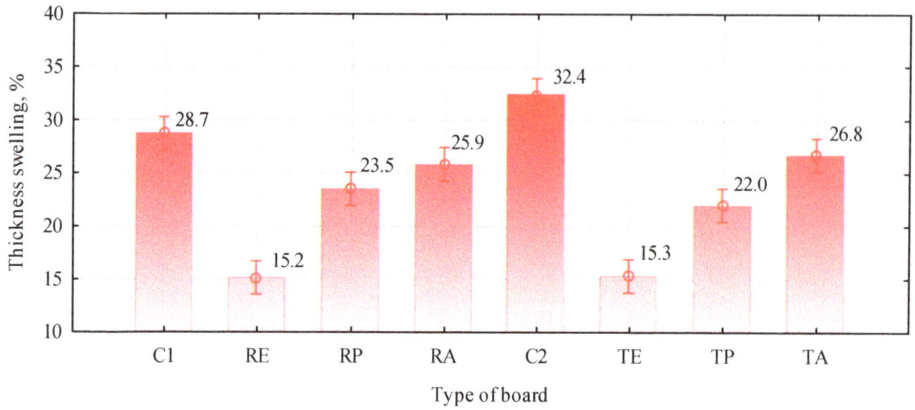

Figure 5. The effect of thermoplastic polymers on the swelling of the thickness of the boards after 24 h of soaking in water, $F(7, 96) = 59.034$, $p = 0.0000$. C1, control with rye straw; C2, control with triticale straw; R, rye straw; T, triticale straw; P, PP; E, HDPE; A, PLA.

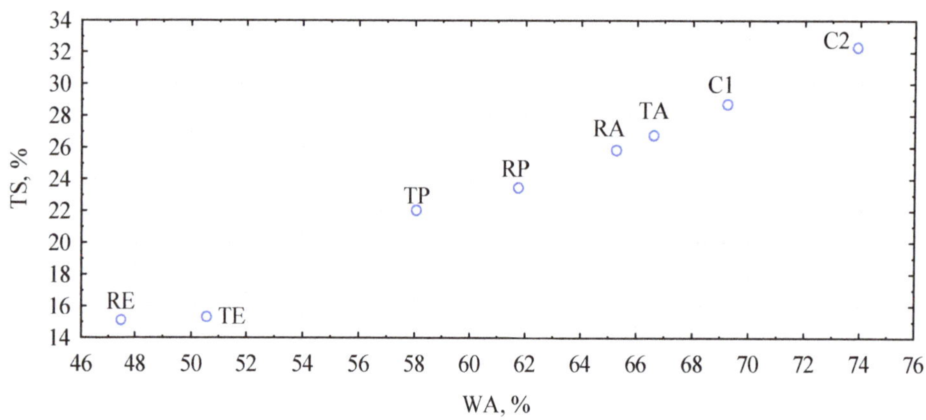

Figure 6. The dependence of board swelling on their water absorption after 24 h of soaking in water. C1, control with rye straw; C2, control with triticale straw; R, rye straw; T, triticale straw; P, PP; E, HDPE; A, PLA.

The assumed board-manufacturing conditions and the type of polymer significantly influenced the formation of the density profile of the produced boards. Figures 7 and 8 show the density profiles for boards made of triticale and rye particles.

Both profiles had a similar shape. Due to the low humidity of the material used to form the outer layers, the zones of maximum density were very strongly shifted into the board's structure. They were located about 4 mm below the surface layers. The middle layer, on the other hand, was compacted to a density of more than 600 kg/m^3. With a low density of near-surface layers, it had lower properties determined in the bend test. The density profile for the RE board looked the opposite way.

In this case, the layers with the highest density were located in the zone of about 1.75 mm from the surface, and the middle layer, from about 6 mm from the surface, had a density close to 500 kg/m^3. Interestingly, the PLA boards had slightly compacted layers at the very surface and areas of maximum density of about 3 mm below the surface. This contributed to the high modulus of elasticity of these boards. The TA boards (triticale with PLA) confirmed this with a profile similar to the control board. Boards with the addition of PP had an intermediate density profile shape. The shape of the profile for any variant was unsymmetrical because the boards were produced in a shelf press. The side on the hot shelf heats up much faster than the top side, which starts heating a few seconds later. In continuous presses, such an effect should not occur. The modification consisting of the production of boards from two species of grass with different bulk densities did not fully bring the expected result. There was some compaction of the outer layers, but it was not as intense as expected. Boards produced as a single layer from the same material under similar conditions had a very flat density profile (Figure 9). Nearly 80% of the thickness of the board had a density of approximately 600 kg/m^3. This adversely affected the static bending strength and, above all, the modulus of elasticity. The higher density of the near-surface layers also delays the thickness swelling of the boards. The introduction of thermoplastic polymers to the outer layers increased the density of these layers; however, an apparent effect was noticed only in the case of RE or RP boards.

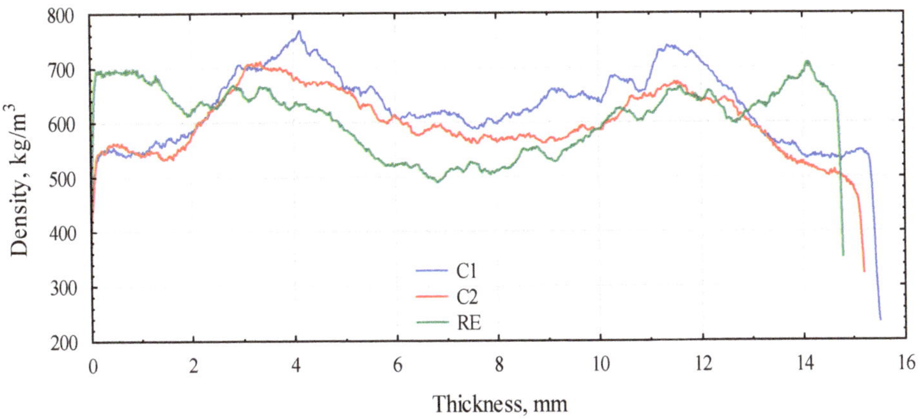

Figure 7. Board density profiles based on triticale and rye HDPE. C1, control with rye straw; C2, control with triticale straw; RE, rye with HDPE.

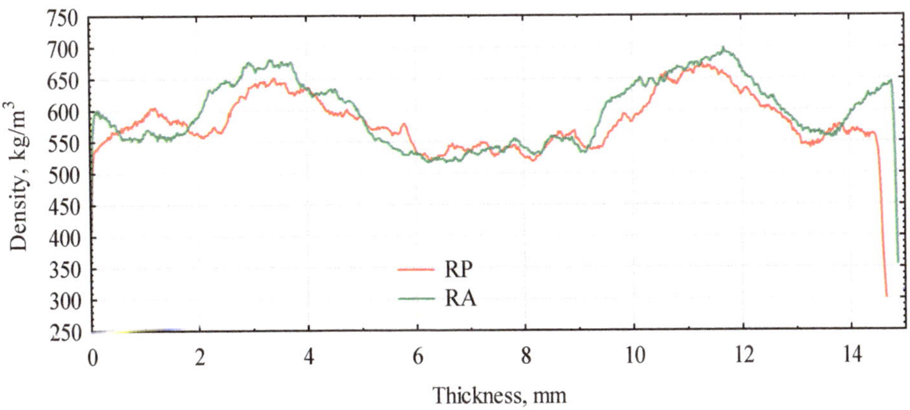

Figure 8. Rye-based board density profiles with PP and PLA. RP, rye straw/PP; RA, rye straw/PLA.

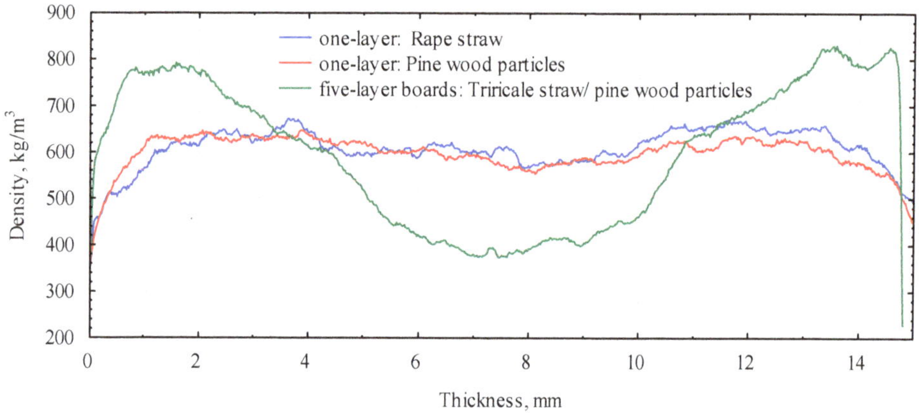

Figure 9. Exemplary density profiles of boards based on straw and wood particles.

However, this was a minor increase compared to boards manufactured as five layers. However, producing three-layer boards with distinctly different layers is much easier than five-layer boards. The shape of the density profile did not significantly affect the tensile strength perpendicular to the surface test (Figure 10).

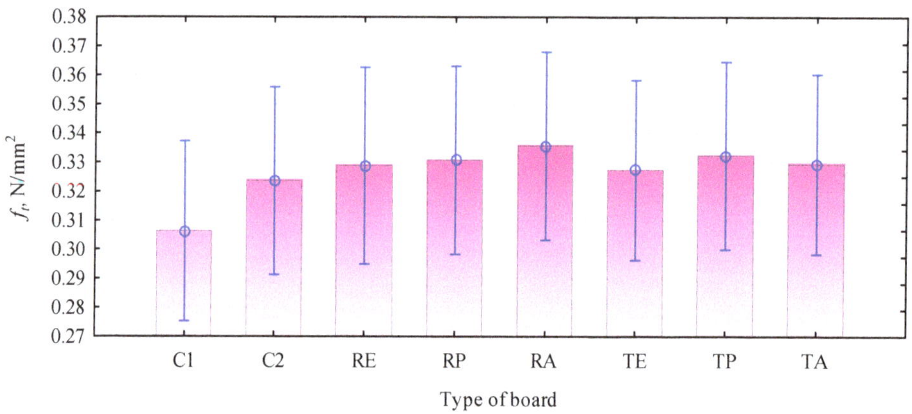

Figure 10. Tensile strength perpendicular to the surfaces of boards based on straw and thermoplastic polymers. C1, control with rye straw; C2, control with triticale straw; R, rye straw; T, triticale straw; P, PP; E, HDPE; A, PLA.

The obtained results ranged from 0.31 N/mm^2 to 0.33 N/mm^2, which was a value close to that often found for this type of plate produced in a laboratory. The samples were destroyed in the middle layer, made of the same straw, and glued similarly; the results should be similar.

Since adding polymers changed the density profile, it could change the damage site to the outer layers. Such an effect, i.e., tearing of the samples in the outer layer, could also have occurred due to the degradation of this layer in the pressing process (wrongly selected pressing conditions and/or incorrect binder is chosen). According to the presented data, the tensile strength perpendicular to the surface test of the analyzed boards did not differ from the previously analyzed rape-based boards of the same density [36]. It was lower by about 20% than the strength of boards with a density of 625 kg/m^3 [11].

Well-prepared pressing conditions allow for the creation of a layer in the outer layer of the boards that blocks water migration into the board. In previous studies, we have shown that when protecting the side surfaces, water penetrated very slowly into the board through the surfaces containing polymers, causing a slight increase in the thickness of the board [31]. Supplementary Figures S1–S3 show the appearance of the surface and its analysis using computed tomography. On the selected cross-sections, the density distribution was determined along the baseline (the green line), while the red line marks the place on the cross-section.

The data in Supplementary Figure S1 refer to the TE board and the structural analysis in the near-surface layer, i.e., about 1 mm below the sample's surface. The surface was covered with a polymer; it was smooth and quite bright. The radiological density defined on the Hounsfield scale ranges from −336 to 336 HU.

The negative value the red point on the profile, is placed in the dark area associated with the surface of the straw, while the positive value is in the white area, which is related to the HDPE. Supplementary Materials Figure S3 shows the profile for the sample's surface in the middle of the board's thickness. The histogram has shown changes from −589 to 25 HU, with no clear high positive values. Air density under standard conditions was defined as −1000 HU, although it was generally assumed to be below −700 HU. The radiation density for wood with a 600–800 kg/m^3 is −400–200 HU [44]. Thus, strands of straw and free spaces dominated the central zone.

In contrast, in the outer zone, in which thermoplastic polymers had bonded with straw chopped, the radiation density value significantly increased above 0 HU. It was confirmed by the radiation density profile on the control sample (board C2), at a similar depth, in the near-surface layer. Similar to the analysis of the middle layer, it was significantly below 0 HU, which indicates the presence of lignocellulosic material with a certain degree of density. The structure of the analyzed straw-based boards was determined by infrared spectroscopy, and the results are presented as spectra in Figure 11. The ATR-FTIR spectra of particleboards made from rye straw contained several well-defined peaks that are characteristic of lignocellulosic materials. The presented spectra were interpreted based on the available literature data [49–53].

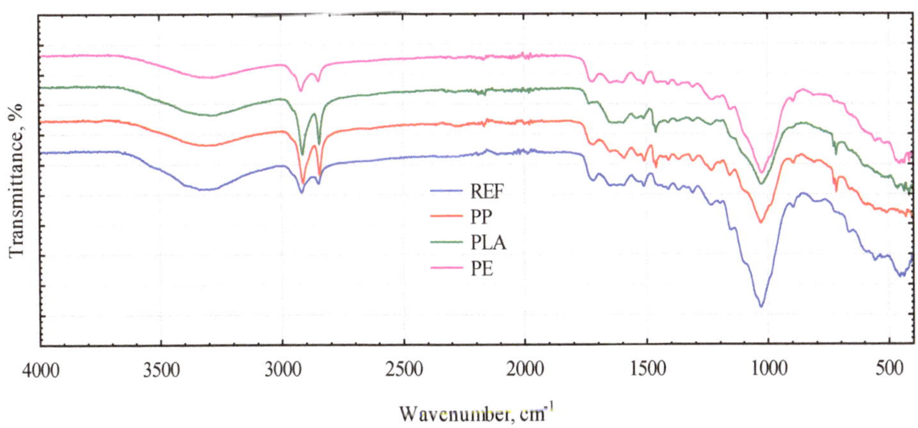

Figure 11. The spectra of particleboards with rye straw. REF, rye straw without polymer; PP, polypropylene; PLA, polylactic acid; PE, high-density polyethylene.

The broad peak between 3600 and 3200 cm^{-1} corresponds to the O-H stretching of water associated with the intermolecular hydrogen bond. The spectrum of the particleboard sample without a polymer matrix showed a higher intensity of this band, which is related to the hydrophilic tendency of lignocellulosic cereal material. The spectra of all samples contain peaks at approximately 2920 cm^{-1} and 2850 cm^{-1}, arising from symmetric and asymmetric stretching vibrations in methoxyl, methyl, and methylene groups in the structure of saccharides and aromatic compounds. The peak at approximately 1720 cm^{-1}

observed in the IR spectra of all samples can be assigned to the C=O stretching of the acetyl ester group of hemicellulose and/or the ester linkage of the carboxylic group of the ferulic and coumaric acids of lignin. The bands in the range of 1600–1500 cm^{-1} can represent the aromatic ring stretch, including the C=C bond of aromatic skeletal vibrations in lignin. The peak at 1460 cm^{-1} can be attributed to the C-H deformation stretching in lignin and xylan. In turn, the 1200–1000 cm^{-1} region represents the C-O stretching and deformation bands in cellulose and lignin, and the bands in the 900–690 cm^{-1} range can be described as C-H bending. In addition, overlapping peaks from the straw and polymer matrix can be observed in the spectra of the polymer-matrix particleboard [54–56]. Similar peaks to those in the particleboard with rye were observed in the spectra of samples with triticale (Figure 12). However, these peaks differ in the intensity of the transmittance, related to the different amounts of characteristic groups (varied content of the main components: cellulose, lignin, and hemicellulose). Moreover, in some cases, the peaks for the same constituent were shifted, which is associated with the nature of hydrogen bonding and coupling effects.

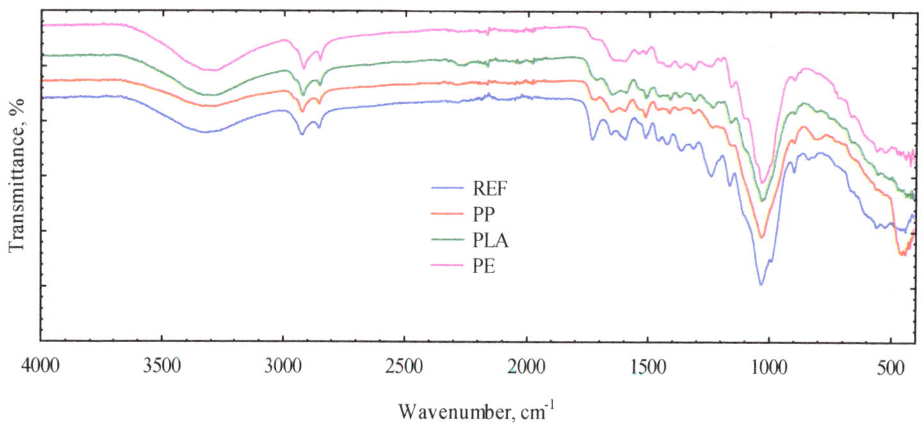

Figure 12. The spectra of particleboards with triticale straw. REF, triticale straw without polymer; PP, polypropylene; PLA, polylactic acid; PE, high-density polyethylene.

The weight losses of triticale-based board samples in relation to the temperature of thermal degradation (TG curve) and the first derivative of that curve (DTG) are presented in Figure 13.

The DTG curve for the sample prepared from triticale without a polymer matrix (Figure 13a) showed a single decomposition step, and the decomposition peak temperature (T_{max}) was 341 °C, which is in agreement with the literature data [38,57]. The boards consisted of triticale, and PLA (Figure 13d) also degraded in one step, just like the samples from the lignocellulosic material alone. The T_{max} of the PLA-triticale composite was observed at the same temperature as for the biomass sample alone and was 341 °C. According to the literature data, the T_{max} for pure PLA is around 362 °C [40]. Therefore, the overlap of decomposition peaks can be observed in this case. Adding another polymer matrix (PP and HDPE) to the biomass caused a change in the thermal behavior of the obtained composites. The boards with PP and HDPE curves (Figure 13b,c) presented two steps for weight loss. The first weight loss of the triticale-PP composite, which can be related to biomass degradation, started at 342 °C. The second weight loss of this composite was at 464 °C and can be connected with the degradation of PP, for which the T_{max}, according to the literature data, is around 455 °C [39]. The two peaks were also observed on the DTG curve of the triticale-HDPE composite, as shown in Figure 13c. The minor degradation peak temperature (T_{max1}) at 340 °C was related to the straw degradation, and the higher degradation peak temperature (T_{max2}) at 467 °C resulted from the degradation of polyethylene. The onset of all composites degradation (T_{onset}) began at 168–189 °C depending on the composite type. According to the literature reports, the thermal degradation of hemicellulose occurs in the temperature range of 150–315 °C, while the depolymerization of cellulose takes place between 275 and 400 °C. In turn, the decomposition of lignin covers a broad temperature range of 150–500 °C [40,41,57]. All composite samples showed a mass loss between 69.59% (triticale composite) and 76.67% (triticale-HDPE composite) upon reaching 500 °C.

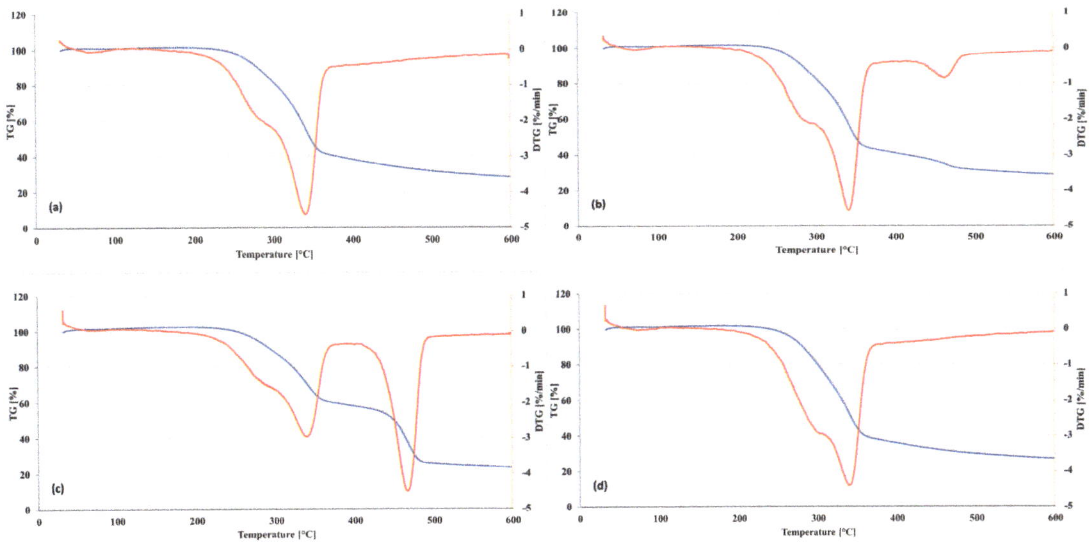

Figure 13. TGA (blue line) and DTG (red line) curves of triticale-based boards: (**a**) control, (**b**) TP, (**c**) TE, and (**d**) TA. T, triticale straw; P, PP; E, HDPE; A, PLA.

4. Conclusions

The properties of boards based on lignocellulose particles, referred to as chips, strips, or chopped strands, depend on several technological factors that have been known for years. However, there is still a need for research to refine specific solutions, especially when trying to add new materials to the production of boards. Based on the result, the following conclusions were drawn:

- There is a possibility of producing lignocellulosic boards from two different straws, the less advantageous of which should be the middle layer of these boards. Because the straws of grass origin stick together more efficiently and form more compact layers, using them in outer layers is better. Rape straws, with a completely different structure, can be successfully used as the middle layer of such boards. The rapeseed straw, with different structures, allows for good shaping of the middle layer's structure, despite the parenchyma tissue's negative impact on the gluing process. However, the research in this area can be continued by even greater diversification of the bulk density or moisture content of the layers;
- Introducing thermoplastics only to the outer layers is beneficial due to the more straightforward melting process. It affected the boards' properties determined in the bend test by affecting the density profile. It can be concluded that the beneficial effect of thermoplastics on the board density profile masks their low elastic properties;
- Thermoplastics, which were successfully merged with the lignocellulosic material, significantly improved the hydrophobic properties of the boards by reducing their swelling in thickness and reducing water absorption;

- Among the straw-based boards with the addition of tested polymers, satisfactory properties were produced mainly using HDPE, and moderate properties using PP. PLA-containing boards did not show clearly favorable properties either in terms of mechanical or physical features;
- The properties of straw–polymer boards produced based on triticale straw were slightly better than rye-based boards, probably due to the geometry of the strands, which was more favorable for triticale straw;
- The particleboards made from straw with the addition of polymers due to lower hygroscopicity and thickness swelling compared to traditional boards can be used in conditions of increased humidity, i.e., garden, bathroom, or kitchen furniture;
- The study should continue to further develop the finishing process.

Supplementary Materials: The following supporting information can be downloaded at: https://www.mdpi.com/article/10.3390/ma16124400/s1, Figure S1. Photos and profile of the density distribution in the HU scale—the analyzed layer at the depth of the red line, the density histogram defined along the green line—TE sample. Figure S2. Photos and profile of the density distribution in the HU scale—the analyzed layer at the depth of the red line (the middle of the sample thickness), the histogram of the density along the green line—the TE sample. Figure S3. Photos and histogram of the density distribution in the HU scale—the analyzed layer at the depth of the red line (the middle of the sample thickness), the histogram of the density along the green line—sample C2.

Author Contributions: Conceptualization A.B. and R.M.; methodology A.B., R.M. and D.D.; formal analysis A.B., R.M. and M.W.; investigation A.B., R.M., D.D. and M.W.; writing—original draft preparation, A.B.; writing—review and editing, A.B., R.M., M.W. and D.D.; supervision, R.M. All authors have read and agreed to the published version of the manuscript.

Funding: The study was also supported by the funding for statutory R&D activities, as research task no. 506.224.02.00 of Faculty of Forestry and Wood Technology, Poznań University of Life Sciences.

Institutional Review Board Statement: Not applicable.

Informed Consent Statement: Not applicable.

Data Availability Statement: Not applicable.

Conflicts of Interest: The authors declare no conflict of interest.

References

1. Mirski, R.; Kawalerczyk, J.; Dziurka, D.; Wieruszewski, M.; Trociński, A. Effects of using bark particles with various dimensions as a filler for urea-formaldehyde resin in plywood. *BioResources* **2020**, *15*, 1692–1701. [CrossRef]
2. Gradziuk, P.; Gradziuk, B.; Trocewicz, A.; Jendrzejewski, B. Potential of straw for energy purposes in Poland—Forecasts based on trend and causal models. *Energies* **2020**, *13*, 5054. [CrossRef]
3. Gradziuk, P. The potential of straw for energy purposes in Poland. *Barom. Reg.* **2014**, *2*, 15–22. [CrossRef]
4. Zhang, Y.; Lu, X.; Pizzi, A.; Delmotte, L. Wheat straw particleboard bonding improvements by enzyme pretreatment. *Holz Als Roh Werkst.* **2003**, *61*, 49–54. [CrossRef]
5. Boquillon, N.; Elbez, G.; Schönfeld, U. Properties of wheat straw particleboards bonded with different types of resin. *J. Wood Sci.* **2004**, *50*, 230–235. [CrossRef]
6. Bekhta, P.; Korkut, S.; Hiziroglu, S. Effect of pretreatment of raw materials on properties of particleboards panels made from wheat straw. *BioResources* **2013**, *8*, 4766–4774. [CrossRef]
7. Azizi, K.; Tabarsa, T.; Ashori, A. Performance characterizations of particleboards made with wheat straw and waste veneer splinters. *Compos. B. Eng.* **2011**, *42*, 2085–2089. [CrossRef]
8. Mo, X.; Cheng, E.; Wang, D.; Sun, X.S. Physical properties of medium-density wheat straw particleboard using different adhesives. *Ind. Crops Prod.* **2003**, *18*, 47–53. [CrossRef]
9. Dziurka, D.; Mirski, R. Lightweight boards from wood and rape straw particles. *Drewno* **2013**, *56*, 19–31.
10. Huang, L.; Xia, P.; Liu, Y.; Fu, Y.; Jiang, Y.; Liu, S.; Wang, X. Production of biodegradable boards using rape straw and analysis of mechanical properties. *BioResources* **2016**, *11*, 772–785. [CrossRef]
11. Mirski, R.; Dziurka, D.; Banaszak, A. Properties of particleboards produced from various lignocellulosic particles. *BioResources* **2018**, *13*, 778–7765. [CrossRef]
12. Li, X.; Cai, Z.; Winandy, J.E.; Basta, A.H. Selected properties of particleboard panels manufactured from rice straws of different geometries. *Bioresour. Technol.* **2010**, *101*, 4662–4666. [CrossRef]

13. Zhang, L.; Hu, Y. Novel lignocellulosic hybrid particleboard composites made from rice straws and coir fibers. *Mater. Des.* **2014**, *55*, 19–26. [CrossRef]
14. Kurochi, Y.; Sato, M. Effect of surface structure, wax silica on the properties of binderless board made from rice straw. *Ind. Crop Prod.* **2015**, *77*, 949–953. [CrossRef]
15. Wu, T.; Wang, X.; Kito, K. Effects of pressures on the mechanical properties of corn straw bio-boards. *Eng. Agri. Environ. Food* **2015**, *8*, 123–129. [CrossRef]
16. Akinyemi, B.A.; Kolajo, T.E.; Adedolu, O. Blended formaldehyde adhesive bonded particleboards made from groundnut shell and rice husk wastes. *Clean Technol. Environ. Policy* **2022**, *24*, 1653–1662. [CrossRef]
17. Dukarska, D.; Bartkowiak, M.; Stachowiak-Wencek, A. White mustard straw as an alternative raw materials in manufacturing particleboards resinated with different amounts of urea formaldehyde resin. *Drewno* **2015**, *58*, 49–63.
18. Dukarska, D.; Łęcka, J.; Czarnecki, R. The effect of wood chip substitution with evening primrose waste on properties of particleboards depending on the type of binding agent. *EJPAU* **2015**, *15*, 5.
19. Bekalo, S.A.; Reinhardt, H.-W. Fibers of coffee husk and hulls for the production of particleboard. *Mater. Struct.* **2010**, *43*, 1049–1060. [CrossRef]
20. Guuntekin, E.; Uner, B.; Karakus, B. Chemical composition of tomato (*Solanum lycopersicum*) stalk and suitability in the particle-board production. *J. Environ. Biol.* **2009**, *30*, 731–734.
21. Nemli, G.; Kırcı, H.; Serdar, B.; Ay, N. Suitability of kiwi (*Actinidia sinensis* Planch.) prunings for particleboard manufacturing. *Ind. Crops Prod.* **2003**, *17*, 39–46. [CrossRef]
22. *EN 310*; Wood-Based Panels—Determination of Modulus of Elasticity in Bending and of Bending Strength. European Committee for Standardization: Brussels, Belgium, 1993.
23. Faustino, J.; Pereira, L.; Soares, S.; Cruz, D.; Paiva, P.; Varum, H.; Ferreira, J.; Pinto, J. Impact sound insulation technique using corn cob particleboard. *Constr. Build Mater.* **2012**, *37*, 153–159. [CrossRef]
24. Pinto, J.; Paiva, A.; Varum, H.; Costa, A.; Cruz, D.; Pereira, S.; Fernandes, L.; Tavares, P.; Agarwal, J. Corn's cob as a potential ecological thermal insulation material. *Energ. Buildings* **2011**, *43*, 1985–1990. [CrossRef]
25. Pinto, J.; Vieira, B.; Pereira, H.; Jacinto, C.; Vilela, P.; Paiva, A.; Pereira, S.; Cunha, V.; Varum, H. Corn cob lightweight concrete for non-structural applications. *Constr. Build. Mater.* **2012**, *34*, 346–351. [CrossRef]
26. Grigoriou, A.H. Straw-wood composites bonded with various adhesive systems. *Wood Sci. Technol.* **2000**, *34*, 355–365. [CrossRef]
27. Cheng, W.; Han, G.; Fang, D. Oriented structural boards from split wheat straw: Effects of straw length, panel density, and resin content. *BioResources* **2013**, *8*, 4497–4504. [CrossRef]
28. Pukszta, D.; Szostak, M.; Rogacz, M. Mechanical properties of polypropylene copolymers composites filled with rapeseed straw. *Polimery* **2014**, *59*, 165–169. [CrossRef]
29. Kuciel, S.; Liber-Kneć, A.; Zajchowski, S. Kompozyty z włóknami naturalnymi na osnowie recyklatu polipropylenu. *Polimery* **2010**, *55*, 718–725. [CrossRef]
30. Sudhakar, K.; Srinivas, C. Investigation of mechanical properties of rice straw fiber polypropylene composites. *Int. J. Eng. Res. Appl.* **2014**, *4*, 182–187.
31. Mirski, R.; Banaszak, A.; Bekhta, P. Selected properties of formaldehyde-free polymer-straw boards made from different types of thermoplastics and different kinds of straw. *Materials* **2021**, *14*, 1216. [CrossRef]
32. Luo, P.; Yang, C.; Li, M.; Wang, Y. Manufacture of thin rice straw particleboards bonded with various polymeric methane diphenyl diisocyanate/ urea formaldehyde resin mixtures. *BioResorces* **2020**, *15*, 935–944. [CrossRef]
33. Pawlicki, J.; Niecewicz, D. Waste thermoplastic polymers in technologies of wood-based panels. In Proceedings of the III International Symposium "Wood Agglomeration", Zvolen, Slovakia, 28–30 June 2020.
34. Borysiuk, P. Możliwość wytwarzania płyt wiórowo-polimerowych z wykorzystaniem poużytkowych termoplastycznych tworzyw sztucznych. *Rozpr. Nauk. Monogr. Szkoła Główna Gospod. Wiej. Warszawie* **2012**, *393*, 1–136.
35. Jastrząb, J. Badania nad Zastosowaniem Termoplastów w Procesie Wytwarzania Płyt OSB. Ph.D. Thesis, Poznań University of Life Science, Poznań, Poland, 2014.
36. Mirski, R.; Dziurka, D.; Banaszak, A. Effects of manufacture conditions on physical and mechanical properties of rape-polymers board. *Wood Res.* **2019**, *64*, 659–666.
37. Mirski, R.; Dziurka, D.; Banaszak, A. Using rape particles in the production of polymer and lignocellulose boards. *BioResources* **2019**, *14*, 6736–6746. [CrossRef]
38. Mihai, M.; Ton-That, M.-T. Valorization of triticale straw biomass as reinforcement in proficient polypropylene biocomposites. *Waste Biomass. Valor.* **2018**, *9*, 1971–1983. [CrossRef]
39. Mihai, M.; Ton-That, M.-T.; Ngo, T.-D.; Busnel, F. New polypropylene/triticale composites: Relationship between formulation and properties. In Proceedings of the 69th Annual Technical Conference of the Society of Plastics Engineers 2011 (ANTEC 2011), Boston, MA, USA, 1–5 May 2011.
40. Mihai, M.; Ton-That, M.-T. Novel polyactide/triticale straw biocomposites: Processing, formulation, and properties. *Polym. Eng. Sci.* **2014**, *54*, 446–458. [CrossRef]
41. Mamun, A.A.; Heim, H.P.; Bledzki, A.K. The use of maize, oat, barley and rye fibers as reinforcements in composites. In *Biofiber Reinforcements in Composites Materials*; Woodhead Publishing: Cambridge, UK, 2015; pp. 454–487.

42. *EN 319*; Particleboards and Fibreboards—Determination of Tensile Strength Perpendicular to the Plane of the Board. European Committee for Standardization: Brussels, Belgium, 1993.
43. *EN-1087-1*; Particleboards—Determination of Moisture Resistance—Boil Test. European Committee for Standardization: Brussels, Belgium, 1995.
44. *EN 317*; Particleboards and Fibreboards. Determination of Swelling in Thickness after Immersion in Water. European Committee for Standardization: Brussels, Belgium, 1993.
45. Dukarska, D.; Czarnecki, R.; Dziurka, D.; Mirski, R. Construction particleboards made from rapeseed straw glued with hybrid pMDI/PF resin. *Eur. J. Wood Wood Prod.* **2017**, *75*, 175–184. [CrossRef]
46. Borysiak, S.; Paukszta, D. Mechanical properties of lignocellulosic/polypropylene composites. *Mol. Cryst. Liq. Cryst.* **2008**, *484*, 13–22. [CrossRef]
47. Borysiuk, P.; Mamiński, M.; Zado, A. Some comments on the manufacturing of thermoplastic-bonded particleboards. *Ann. Wars. Univ. Life Sci. For. Wood Technol.* **2009**, *68*, 50–54.
48. Borysiuk, P.; Wilkowski, J.; Krajewski, K.; Auriga, R.; Skomorucha, A.; Auriga, A. Selected properties of flat-pressed wood-polymer composites for high humidity condition. *BioResources* **2020**, *15*, 5141–5155. [CrossRef]
49. Bledzki, A.K.; Mamun, A.A.; Volk, J. Physical, chemical and surface properties of wheat husk, rye husk and soft wood and their polypropylene composites. *Compos. Part A* **2010**, *41*, 480–488. [CrossRef]
50. Sun, R.C.; Sun, X.F. Structural and thermal characterization of acetylated rice, wheat, rye, and barley straws and poplar wood fibre. *Ind. Crops Prod.* **2002**, *16*, 225–235. [CrossRef]
51. Kalina, M.; Sovova, S.; Svec, J.; Trudicova, M.; Hajzler, J.; Kubikova, L.; Enev, V. The effect of pyrolysis temperature and the source biomass on the properties of biochar produced for the agronomical applications as the soil conditioner. *Materials* **2022**, *15*, 8855. [CrossRef]
52. Vitolina, S.; Shulga, G.; Neiberte, G.; Jaunslavietis, J.; Verovkins, A.; Betkers, T. Characteristics of the waste wood biomass and its effect on the properties of wood sanding dust/recycled PP composite. *Polymers* **2022**, *14*, 468. [CrossRef] [PubMed]
53. Woźniak, M.; Ratajczak, I.; Wojcieszak, D.; Waśkiewicz, A.; Szentner, K.; Przybył, J.; Borysiak, S.; Goliński, P. Chemical and structural characterization of maize stover fractions in aspect of its possible applications. *Materials* **2021**, *14*, 1527. [CrossRef] [PubMed]
54. Luna, I.Z.; Dam, K.C.; Chowdhury, S.; Gafur, A.; Khan, N.; Khan, R.A. Physical and thermal characterization of alkali treated rice husk reinforced polypropylene composites. *Adv. Mater. Sci. Eng.* **2015**, *2015*, 907327. [CrossRef]
55. Mofokeng, J.P.; Luyt, A.S.; Tabi, T.; Kovacs, J. Comparison of injection moulded, natural fibre-reinforced composites with PP and PLA as matrices. *J. Thermoplast. Compos. Mater.* **2012**, *25*, 927–948. [CrossRef]
56. Singla, P.; Mehta, R.; Berek, D.; Upadhyay, S.N. Microwave assisted synthesis of poly(lactic acid) and its characterization using size exclusion chromatography. *J. Macromol. Sci.* **2012**, *49*, 963–970. [CrossRef]
57. Ross, K.; Godfrey, D. Effect of extractives on the thermal decomposition of wheat, triticale and flax crop residues: A kinetic study. *IJBR* **2012**, *1*, 19–31.

materials

MDPI

Article

Flake Graphene as an Innovative Additive to Grease with Improved Tribological Properties

Małgorzata Djas [1,2,*], Anna Matuszewska [3], Beata Borowa [3], Krystian Kowiorski [1], Piotr Wieczorek [3], Marcin Małek [4] and Adrian Chlanda [1]

1 Łukasiewicz Research Network—Institute of Microelectronics and Photonics, al. Lotników 32/46, 02-668 Warsaw, Poland
2 Faculty of Chemical and Process Engineering, Warsaw University of Technology, Waryńskiego 1, 02-645 Warsaw, Poland
3 Łukasiewicz Research Network—Automotive Industry Institute, Jagiellońska 55, 03-301 Warsaw, Poland
4 Faculty of Civil Engineering and Geodesy, Military University of Technology, Gen. Sylwestra Kaliskiego 2, 00-908 Warsaw, Poland
* Correspondence: malgorzata.djas@imif.lukasiewicz.gov.pl

Abstract: The paper presents the results of research on the use of flake graphene as an additive to plastic grease in order to improve its tribological properties. The influence of concentration (0.25–5.00 wt.%) and the form of graphene (graphene oxide, reduced graphene oxide) on selected properties of the base grease were investigated. It has been found that the addition of graphene flakes improves the anti-wear properties of the lubricant. The greatest improvement in the properties of the lubricant was achieved by using graphene at a concentration of 4.00 wt.%; the reduction in the average diameter of the wear scar was almost 70% for GO and RGO, compared to the base lubricant without the addition of graphene.

Keywords: flake graphene; graphene oxide; reduced graphene oxide; lubricants; grease

Citation: Djas, M.; Matuszewska, A.; Borowa, B.; Kowiorski, K.; Wieczorek, P.; Małek, M.; Chlanda, A. Flake Graphene as an Innovative Additive to Grease with Improved Tribological Properties. *Materials* **2022**, *15*, 7775. https://doi.org/10.3390/ma15217775

Academic Editor: Sergio Morales-Torres

Received: 6 October 2022
Accepted: 28 October 2022
Published: 4 November 2022

1. Introduction

Lubricants, such as oils and greases play a key role in many industries, including the automotive, transportation, chemical and food industries in particular. The factor driving the growth of the market for these products in the world is the dynamic technological development. Lubricant manufacturers have to keep pace with more and more advanced vehicles as well as innovative machines and devices, where precision and smooth operation are crucial for their efficient and profitable operation. During mechanical processes, as a result of strong friction between the surfaces, wear of the mechanical elements of the friction junction occurs and heat is generated, which ultimately reduces the efficiency of work and shortens the life of the device. It is estimated that about 23% (119 EJ) of global energy consumption is due to anti-friction and regeneration of friction elements [1]. The reduction of friction not only contributes to the reduction of energy consumption, but also fuel savings and the related financial and environmental benefits, in particular the reduction of CO_2 emissions [2]. In addition, conventional additives with anti-wear (AW) and anti-seizure (extreme pressure EP) properties are mainly based on sulfur, chlorine or phosphorus compounds, which are not environmentally friendly [3,4]. Therefore, it is necessary to develop innovative components, modifiers and additives to lubricants with significantly improved tribological properties and reduced negative impact on the natural environment. It has been found that one such additive may be flake graphene, including its derivatives (graphene oxide—GO; reduced graphene oxide—RGO), due to its unique properties, such as thickness of only one or a few carbon layers, layered structure, small flake size. Having in mind flake graphene materials' thickness, most of them are classified as nanoadditives. Conventional EP and AW additives act by adsorbing active groups to the friction surface or react chemically with the metal surface to form a modified boundary layer. On the other hand, EP and AW nanoadditives can protect frictional mating surfaces mainly through: mending effect, polishing effect, protective film formation mechanism and rolling bearing effect [5,6].

Flake graphene, which is a known 2D carbon nanomaterial, is obtained from graphite. At the same time, it should be emphasized that graphite is characterized with greasing properties and it is widely implemented, i.e., as an anti-wear agent. Graphene is another allotropic type of carbon, next to graphite. GO is the oxidized form of graphene, decorated with oxygen functional groups (e.g., hydroxyl, carboxyl or alkoxy groups). GO particles are highly hydrophilic and form stable dispersions in water and some organic solvents. RGO is obtained by reduction of oxygen functional groups of GO. Contrary to GO, rGO is hydrophobic and show high mechanical strength, large specific surface area and high electric and thermal conductivity [7].

The use of flake graphene is currently the subject of much scientific research in various areas, such as composites [8], functional coatings [9], electronics [10], energy storage [11], sensors [12] and even medicine and tissue engineering [13,14]. One of the intensively developed and promising areas of application of flake graphene is tribology. It was found that the introduction of flake graphene as an additive to the lubricant has a positive effect on its tribological properties, reducing the value of the friction coefficient and the degree of surface wear. Better lubricity is achieved by nanostructured arrangement of graphene flakes and even by graphitization. In addition, the presence of a protective film on the surface of the friction junction elements, arranged parallel to the direction of movement, was found, which indicates that the graphene flakes slide between each other due to friction [15].

As part of the research described in the literature, the tribological properties of various lubricants (lubricating oils, plastic greases) with the addition of various forms of graphene flakes (graphene, graphene oxide, reduced graphene oxide, graphene nanoplatelets) were analyzed. The tribological properties of polyalphaolefin-2 (PAO2) base oil with the addition of multilayer graphene containing 0.05 wt.%, 0.10 wt.% and 0.50 wt.% were investigated using a 4-ball tester. The tests were carried out at loads from 120 N to 400 N and rotational speeds in the range of 100–400 rpm [16]. It was found that, regardless of the tested concentration, the use of graphene in the oil improves the tribological properties (reduction of the friction coefficient) of the oil compared to the oil without the additive, with the best results being obtained for a concentration of 0.05 wt.%. However, in the case of a 10-fold increase in the concentration of graphene in the oil to 0.50 wt.%, the value of the friction coefficient increased, approaching the results obtained for pure oil [16]. The reason for this may be the tendency of the graphene material to agglomerate in the base oils. To prevent this from happening, dispersants can be added to the oil or the graphene can be chemically modified [17,18].

Graphene as an anti-wear additive to plastic greases was also tested. The properties of mineral plastic grease thickened with lithium saponified, enriched with the addition of graphene nanoplatelets with a concentration of 0.50 wt.%, 1.00 wt.%, 5.00 wt.% and 10.00 wt.% were investigated. Various loads were applied: 5 N, 10 N and 15 N [19]. It was found that with an increase in the concentration of graphene in the lubricant, the value of the friction coefficient decreased, in contrast to the results presented earlier, where the best results were obtained for samples containing the lowest concentrations of additives. The value of the friction coefficient decreased by 8.5% for the concentration of 0.50 wt.%, while the increase in the concentration of graphene to 10.00 wt.% resulted in a reduction of the friction coefficient by 16.3% compared to the value of the coefficient obtained for pure lubricant [19]. Fu et al. found that the addition of graphene nanoplatelets to the base grease effectively reduces the value of the friction coefficient and improves the thermal conductivity of the lubricant. The concentration of graphene in the lubricant was 1.00–4.00 wt.%. The best results of the lubricant's tribological properties were obtained for the graphene concentration in the lubricant of 2.00 wt.% [20], for which the diameter of the friction scar decreased by approx. 14% compared to the base grease.

Flake graphene offers great research opportunities because its surface can be chemically modified and its properties can be controlled in this way. For example, the lubricating properties of monolayer graphene oxide when used as an additive to water-based lubricants were investigated [21]. Tribological tests were performed using a reciprocating sliding configuration at node load 1.88 N. It was found that the addition of GO to the water improved lubrication and provided a very low coefficient of friction of about 0.05 with no apparent wear to the surface after 60,000 cycles of friction tests. It was concluded that the adsorption of graphene oxide on the lubricated surfaces of the friction elements is responsible for the protection of the surface and the material behaves like a protective coating. The change in the composition of the graphene surface, in particular the share of oxygen functional groups, also influences the change in the tribological properties of the graphene material [21]. Wang et al. discovered that in the case of the oxidized form of graphene, the friction between the two surfaces increases with the introduction of epoxy and hydroxyl groups into the molecules. The interaction between the layers is dominated by hydrogen bonds [22]. Liu et al. investigated the effect of two types of graphene, physical graphene produced by liquid-phase exfoliation and graphene produced by chemical oxidation and reduction, on the tribological properties of the oil. It was found that the addition of reduced graphene oxide results in worse anti-friction and anti-wear properties of the oil due to the presence of oxygen functional groups on the surface, compared to physical graphene without the presence of oxygen groups. The inferior lubricating properties of RGO are explained by the presence of oxygen-containing groups that increase the surface activity of the flakes, as well as many defects in the material and the contact surface [23]. Another study found that the lubricating properties of the base oil also depend on the number of layers and the size of the RGO particles. It was shown that the best tribological properties were obtained for graphene with a smaller flake size and containing a larger number of layers; the reduction of friction was 37% and wear was 47% lower compared to pure base oil [24].

Based on the literature review, it can be concluded that the addition of graphene to lubricants improves their tribological properties. However, graphene still remains the object of tribological research and the obtained results are not always unequivocal. This may result from the use of various forms of graphene (graphene oxide, reduced graphene oxide, graphene nanoplatelets), different sizes of flakes, different number of carbon layers, different content of oxygen functional groups on the surface and different media in which graphene and its derivatives are dispersed, such as also different test conditions. Therefore, the available information is still incomplete and requires further research. The complexity of the application of graphene in terms of improving the tribological properties of lubricants is also indicated by Larrson et al. [25]. Moreover, in the source literature there are significant discrepancies in the results of the research concerning the influence of graphene concentration in the base lubricant on its tribological properties. The literature lacks results of comparative studies on the effect of the type of graphene (e.g., graphene oxide, reduced graphene oxide, graphene nanoplatelets) on the properties of the lubricant. A very important issue is also the development of a method of introducing graphene into grease or oil in order to obtain a homogeneous dispersion of the nanofiller in the lubricant and, consequently, to improve the lubricity of the product. The literature discusses the problem of agglomeration of graphene flakes; in particular, this problem concerns lubricating oils [20].

Bearing in mind the above-mentioned limitations of the available literature data, this work proposes an innovative approach to the implementation of graphene materials as a nanoadditive with anti-wear properties to greases. For the purposes of the work, graphene materials were designed and manufactured, representing two types of flake graphene: graphene oxide and reduced graphene oxide. The paper presents the comparative results obtained for both materials. In order to obtain the best knowledge about the influence of graphene materials on the anti-wear properties of the grease, the tests were carried out for a wide range of concentrations of graphene materials in the grease, amounting to 0.25–5.00 wt.%. The experiment designed in this way allowed to obtain a detailed quantitative description of the issue under study. In addition, the article attempts to answer which of the two tested types of graphene materials will better meet the requirements for carbon nanoadditives for lubricants, taking into account their practical application.

2. Materials and Methods

The research was conducted using graphene oxide powder and reduced graphene oxide powder by G-Flake® (Łukasiewicz Research Network—Institute of Microelectronics and Photonics, Poland)—Figure 1.

Figure 1. Graphene materials used in the tests: (**a**) graphene oxide, (**b**) reduced graphene oxide.

Graphene oxide was produced using a modified Hummers method. This method is based on the oxidation of flake graphite (Asbury Carbons, Asbury, NJ, USA) in sulfuric acid VI (96%, pure p.a., Chempur, Poland) with the addition of potassium permanganate (pure p.a., Chempur, Piekary Śląskie, Poland) and potassium nitrate (pure p.a., Chempur, Poland). After the oxidation process, the graphite oxide in the form of an aqueous suspension is exfoliated and then purified by filtration. After the purification process, an aqueous suspension of graphene oxide is obtained. In order to produce reduced graphene oxide, graphene oxide undergoes a reduction reaction, which takes place at high a temperature in the presence of a reducing agent—hydrazine (pure p.a., Chempur, Piekary Śląskie, Poland). After the reduction process, the reduced graphene oxide is purified by filtration. To obtain GO or RGO in powder form, the aqueous material dispersion was dried. The obtained graphene materials (GO and RGO) were then characterized using the following methods: scanning electron microscopy (SEM) (Auriga® CrossBeam® Workstation, Carl Zeiss, Oberkochen, Germany, at a voltage of 1 kV, energy selective backscattered EsB detector), Raman spectroscopy (Renishaw Invia Raman Microscope, Wotton-under-Edge, UK, a laser with a wavelength of 532 nm was used for excitation), elemental analysis (628 Series and 836 Series, Leco Corporation, St. Joseph, MI, USA). Regarding the SEM experiment, it was implemented in order to visualize GO and RGO materials. The powders were placed on a silicon wafer and then glued with a carbon tape to an SEM table. No coating with conductive agent was conducted prior to visualization. As for the Raman examination, collected raw data were analyzed with Wire 3.54 software provided by the spectrometer producer. The analysis was carried out in order to discriminate the position of D and G peaks using a Gaussian/Lorentzian. Prior to determination of the peaks, position baseline subtraction was implemented. Detailed characterization of the purified GO water suspension, from which GO powder and RGO powder were then obtained, is presented in our earlier publication [26].

The tests of the lubricating properties of GO and RGO were carried out with the use of plastic grease. The plastic base lubricant was prepared using Finavestan A 360 B paraffin oil (pure, TOTAL, Gonfreville-l'Orcher, France) and lithium stearate (pure, ROTH, Karlsruhe, Germany) thickener. It was prepared in laboratory conditions by adding the thickener lithium stearate to the paraffinic base oil, mixing and then heating with constant stirring until the thickener completely dissolved (to approx. 220 °C). The dissolved mixture was allowed to cool completely at room temperature.

The lubricant compositions were prepared in laboratory conditions by adding the appropriate amount of GO or RGO in the form of a powder to the lubricant base. The samples were then mixed using a planetary mixer (SpeedMixer DAC 150.1 FV-K, Hauschild Engineering, Hamm, Germany) to obtain a homogeneous dispersion (3500 rpm, 15 min). Mixtures of grease and flake graphene were subjected to tribological tests, in which the concentration of graphene (GO, RGO) was from 0.25 wt.% up to 5.00 wt.%. It was observed that as a result of the shear stresses arising when mixing pure base grease and base grease with the addition of graphene, the viscosity of the grease decreased due to the shear-thinning phenomenon (Figure 2).

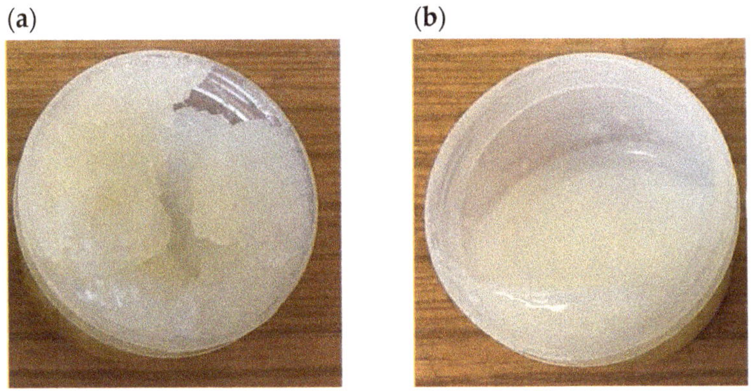

Figure 2. The appearance of the base grease: (**a**) before mixing; (**b**) after mixing in a planetary mixer.

The authors of the study [19] described analogous observations. The authors of this article have experimentally verified that the occurrence of this phenomenon does not affect the properties of the base lubricant and the obtained test results. On the other hand, the high shear stresses arising during the preparation of the mixtures prevent the agglomeration of graphene; therefore, it was possible to obtain a homogeneous dispersion of grease and graphene.

The tests of the lubricating properties of the obtained plastic greases with the addition of flake graphene were determined by measuring the average diameter of the wear scar. The tests were performed with the use of a 4-ball tester (Stanhope-SETA Limited, Chertsey, UK). In this tester, the elements of the tribological couple are in point contact and a sliding movement is performed. The tribological couple in the tester consists of four balls with a diameter of 12.7 mm, made of bearing steel with a hardness of 62.7 HRC. Three balls (stationary) are placed in the lower holder, into which the tested lubricant is also introduced (in the amount of 8 ± 2 cm^3). These balls are pressed down with a ring and fixed with a tightened clamping nut. The fourth ball (movable) is mounted in the upper holder. During the test, it rotates at a speed of 1450 ± 50 rpm, which corresponds to the speed of the rubbing surfaces of 0.55 $m \cdot s^{-1}$. The balls in the lower holder are pressed with the force P against the ball mounted in the upper holder by means of a lever. The friction linkage diagram is shown in Figure 3.

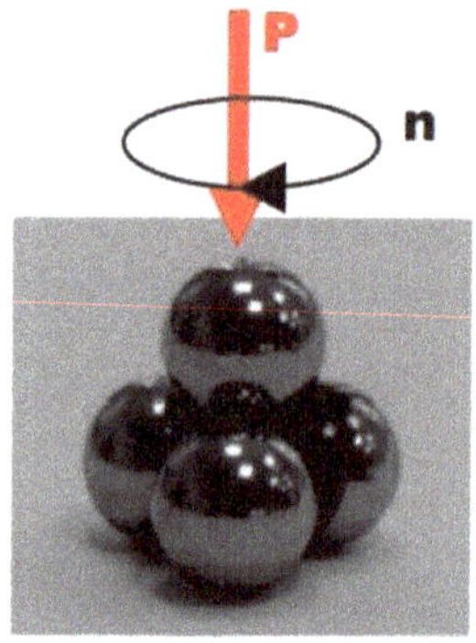

Figure 3. Diagram of a friction linkage in a 4-ball tester.

Before starting the tests, the balls were washed in gasoline and, after drying, stored in acetone. At the end of each run, the stationary balls were washed in extractive gasoline. At least three test cycles were carried out for each lubricant. After the tribological tests, the diameter of the friction scars formed on the test elements was measured. The diameter was measured on three balls from a given run, perpendicular (d_1) and parallel (d_2) to the friction scars, and then the arithmetic mean ($\overline{d}$) was calculated, which was taken as the mean track diameter in a given test run. The principle of calculating the average diameter of the wear scar is shown in Figure 4 and Equation (1).

$$\overline{d} = \frac{d_1^1 + d_2^1 + d_1^2 + d_2^2 + d_1^3 + d_2^3}{6} \quad (1)$$

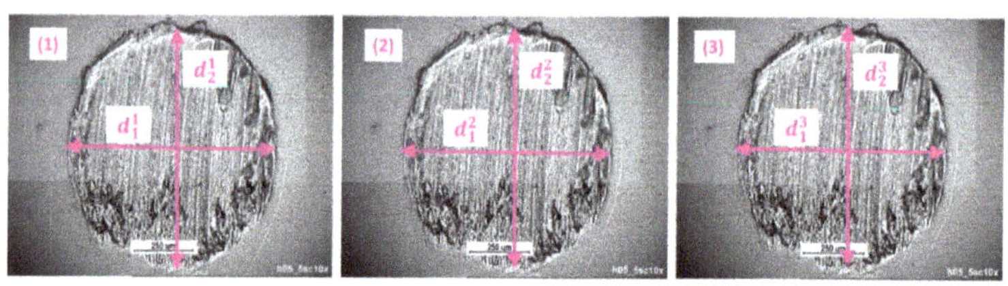

Figure 4. The principle of calculating the average diameter of the wear scar in a given test run: (**1**) the wear scar on the first ball, (**2**) the wear scar on the second ball, (**3**) the wear scar on the third ball.

The size of the trace diameter was the basic measure of the tribological properties of the tested substances. The arithmetic mean of the results of at least three determinations which did not differ from their arithmetic mean by more than 10% was taken as the result of the determination. During the research work with the use of a 4-ball tester, two types of tests were carried out: at a load of 60 kgf for 60 min; at a load of 100 kgf for 1 min. The adopted conditions result from the internal procedure for testing anti-wear additives.

3. Results and Discussion

3.1. Characterization of Flake Graphene

Figure 5 shows SEM images of graphene oxide and reduced graphene oxide. Figure 5a shows GO in the form of a powder, the size of the structures is 5–30 μm, the flake size of GO in the aqueous suspension from which the powder was obtained is 2–10 μm [26] (drying causes agglomerates of graphene flakes). The SEM image of the RGO powder is shown in Figure 5b; the size of the structures is 5–20 μm.

Figure 5. SEM images of (**a**) graphene oxide, (**b**) reduced graphene oxide.

Figure 6 shows the Raman spectra for both tested graphene derivatives GO and RGO. Raman spectroscopy was used to estimate the degree of reduction of GO and to define structural differences between GO and RGO. Raman spectroscopy has been proven to be useful for detailed research description of graphene materials [27,28]. The Raman spectrum of carbon materials contains bands marked as D and G. The D band (located near 1350 cm^{-1}) results from the presence of vacancies or dislocations in the graphene layer and at the edge of this layer. This band is also related to the presence of defects in the material. The G band (located near 1590 cm^{-1}) corresponds to the sp^2 hybridization of the carbon network and is attributed to the first-order scattering from the doubly degenerate E2g phonon modes of graphite in the Brillouin zone center as well as bond stretching of sp^2 carbon pairs in both rings and chains [29]. The ID/IG ratio is related to the amount of defects present in the material [30]. The analysis of Raman spectra for both materials (Figure 6)—GO and RGO—indicates the presence of an intense peak D at a length of ~1357 cm^{-1} for GO and ~1347 cm^{-1} for RGO, respectively, and a G peak at ~1586 cm^{-1}. The ID/IG ratio for GO is 1.05 and for RGO 1.20. The value of the ID/IG parameter increases with the increase in the degree of disorder in the structure of the graphene material.

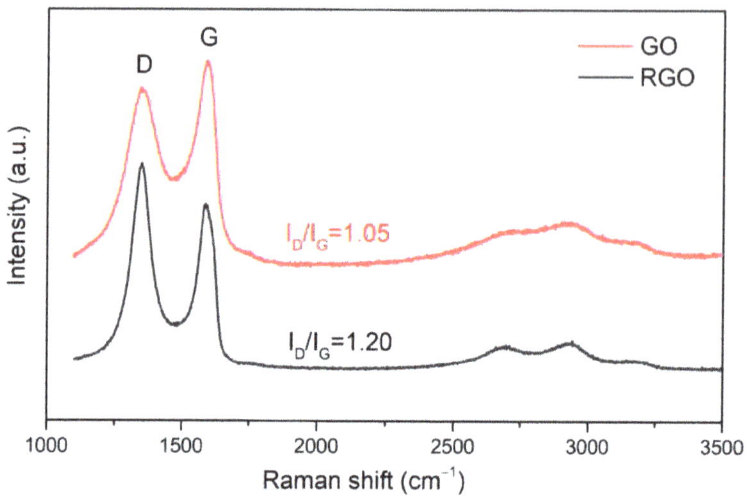

Figure 6. Raman spectra of graphene oxide and reduced graphene oxide.

The results of elemental analysis of graphene materials are presented in Table 1. The oxygen content in GO is 45.0–52.0%, while in the case of RGO it is 15.0–20.0%. The carbon content is 40.0–42.0% for GO and 70.0–80.0% for RGO. The content of the other elements—sulfur, nitrogen and hydrogen—is less than 3.0%.

Table 1. Elemental analysis of graphene oxide and reduced graphene oxide.

	Composition (%)				
	C	O	S	N	H
GO	40.0–42.0	45.0–52.0	1.0–3.0	<0.3	2.0–3.0
RGO	70.0–80.0	15.0–20.0	<2.0	<0.3	<0.2

3.2. *Anti-Wear Properties of the Grease*

The grease is a colloidal system, where the dispersing phase is oil and the dispersed phase is thickeners. The role of the thickener is to ensure the proper spatial structure of the plastic grease and the specific, required rheological and tribological properties. The spatial structure formed in the lubricant by the thickener fibers, as a result of the action of shear forces and the temperature increase during the operation of the tribological node, disintegrates [31,32]. An example of the appearance of the base grease in the lower holder of the 4-ball tester before and after the test (load 100 kgf, time 1 min) is shown in Figure 7. The lubricant in the mating contact changes its structure from semi-liquid (Figure 7a) to liquid (Figure 7b) as a result of shear forces and an increase in temperature in the contact zone of friction elements. The grease is non-Newtonian liquid, classified as thixotropic liquid. Under the influence of shear forces, its internal structure is destroyed and viscosity is reduced [33]. Oil is separated as a result of breaking the bonds of the cross-linker, which enables better lubrication.

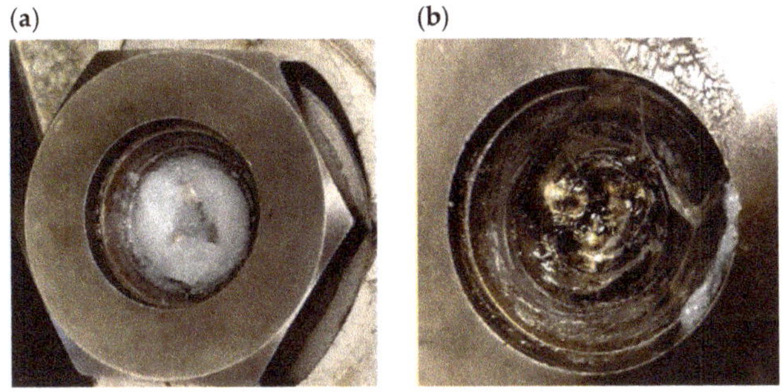

Figure 7. Base grease in the lower handle of the 4-ball tester: (**a**) before and (**b**) after the tribological test-node load 100 kgf for 1 min.

Prepared lubricants with different contents of graphene material (GO, RGO) were subjected to anti-wear tests. Figure 8 shows the results of studies on the effect of GO concentration in the range of 0.25–5.00 wt.%. on lubricating properties, for a load of 60 kgf for 60 min. The obtained data show that the introduction of GO to the base lubricant in a small concentration of 0.25 wt.%. and 0.50 wt.%, increases the wear of the test pieces by approx. 37% (4.48 ± 0.20 mm) and 24% (4.04 ± 0.36 mm), respectively, in relation to the base lubricant without additive (3.26 ± 0.28 mm). However, the introduction of a GO additive in an amount of 1.00 wt.% results in a significant improvement in lubricating properties and reduction of wear. Then, as the concentration of GO in the lubricant increases from 1.00 wt.% to 4.00 wt.%, the degree of reduction of the average wear scar diameter increases from 22% (2.53 ± 0.07 mm) to 69% (1.00 ± 0.05 mm). A reduction value of the average wear scar diameter for a concentration of 3.00 wt.% was similar to 1.00 wt.%, amounting to approx. 24%. Further increase of the GO concentration to 5.00 wt.% results in a re-increase in the average diameter of the wear scar, but still the degree of wear is 40% less compared to the base grease

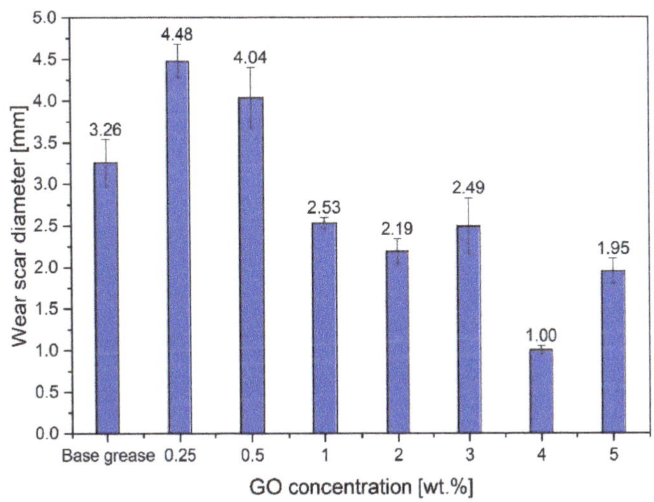

Figure 8. Influence of the concentration of GO additive on the tribological properties of a lubricant.

Figure 9 shows the results of the research on the effect of the concentration of the second form of grapheme—RGO (0.25–5.00%)—on the lubricating properties, for a load of 60 kgf for 60 min.

100 kgf is 27% (2.17 ± 0.07 mm) and at 60 kgf as much as 69% (1.01 mm ± 0.05), compared to pure base grease. This may indicate the anti-wear rather than anti-seizure character of the protective layers created by the nanoadditive.

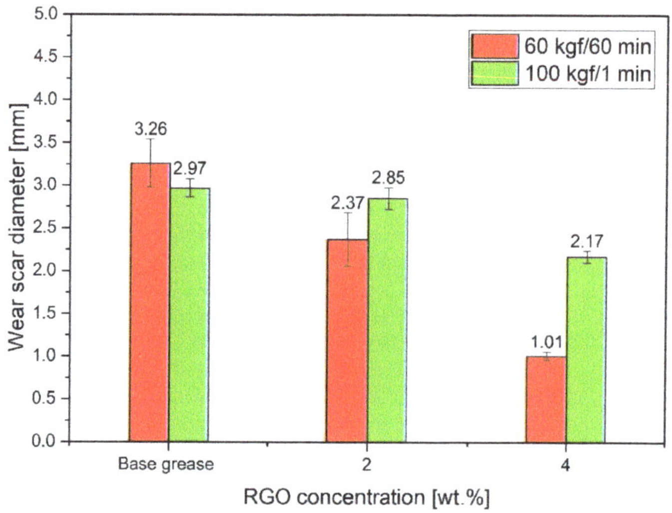

Figure 10. Average diameter of the wear scar of grease with RGO additive depending on the load (60 kgf for 60 min and 100 kgf for 1 min).

4. Conclusions

Grease samples were produced with the addition of GO and RGO with a concentration in the range of 0.25–5.00 wt.%. It was found that the addition of flake graphene improves the anti-wear properties of the lubricant, while the concentration and type of graphene have an impact on the tested lubricant properties and its stability. For lower contents of graphene additive (0.25 wt.% and 0.50 wt.% for GO and 0.25 wt.% for RGO), an increase in wear is observed as compared to the base grease. For low RGO concentrations (0.50 wt.%), the grease properties are similar to those of the base grease without additive. The introduction of flake graphene to the base of the plastic grease may disturb the internal network of the grease due to the high activity of the surface of the graphene flakes. This phenomenon may be the cause of increased node wear at low additive concentrations (0.25 wt.% and 0.50 wt.%), as these contents are probably too low to compensate for the breakdown of the lubricant's inner network. This effect is particularly evident in GO containing numerous oxygen functional groups on the surface of the flakes.

The use of a higher concentration of graphene 1.00–5.00 wt.% results in a reduction in the average diameter of the wear scar, which is greatest for an additive content of 4.00 wt.% (reduction of the wear scar diameter was 69% for GO, as well as for RGO). A further increase in the concentration of the additive to 5.00 wt.% results in a renewed increase in wear, which may result from the agglomeration of the graphene material and a significant increase in viscosity, so that the lubricant is not able to produce a lubricating film that will protect the friction surface well.

Out of the two tested forms of flake graphene, RGO is a recommended additive for lubricants due to its stability under operating conditions, while GO was reduced during friction. The RGO powder tested in operation can therefore be successfully used as a material improving the tribological properties of lubricants.

The use of flake graphene in lubricants is not a thoroughly researched and explained issue. In order to further develop the application of graphene flakes as a lubricant additive, it is necessary to continue research, including in the direction of chemical modification of the flake surface, and to explain the mechanisms of graphene operation during friction, and also to maintain the properties of lubricants unchanged in subsequent work cycles.

Author Contributions: Conceptualization, M.D. and A.M.; methodology, M.D., A.M. and B.B.; validation, M.D. and A.M.; formal analysis M.D., A.M., P.W. and A.C.; investigation, M.D, A.M., B.B., K.K., P.W., M.M. and A.C.; resources, M.D., A.M. and M.M.; data curation, M.D. and A.M.; writing—original draft preparation, M.D, A.M., K.K., P.W., M.M. and A.C.; writing—review and editing, M.D., A.M. and P.W.; visualization, M.D. and A.M.; supervision, M.D and A.M.; project administration, M.D. and A.M.; funding acquisition, M.D., A.M. and M.M. All authors have read and agreed to the published version of the manuscript.

Funding: This research was funded by Łukasiewicz Research Network—Institute of Microelectronics and Photonics and Łukasiewicz Research Network—Automotive Industry Institute (Statutory Research 2020). In addition, this research was funded by National Science Centre, Poland—grant "Identification of mechanisms and investigations of flake graphene production by direct exfoliation using supercritical carbon dioxide", project number 2019/35/D/ST8/02977.

Institutional Review Board Statement: Not applicable.

Informed Consent Statement: Not applicable.

Data Availability Statement: Data are contained within the article.

Conflicts of Interest: The authors declare no conflict of interest.

References

1. Holmberg, K.; Erdemir, A. Influence of tribology on global energy consumption, costs and emissions. *Friction* **2017**, *5*, 263–284. [CrossRef]
2. Liu, L.; Zhou, M.; Jin, L.; Li, L.; Mo, Y.; Su, G.; Li, X.; Zhu, H.; Tian, Y. Recent advances in friction and lubrication of graphene and other 2D materials: Mechanisms and applications. *Friction* **2019**, *7*, 199–216. [CrossRef]
3. Matuszewska, A.; Gradkowski, M. Antiwear Action of Mineral Lubricants Modified by Conventional and Uncoventional Additives. *Tribol. Lett.* **2007**, *27*, 177–180. [CrossRef]
4. Waara, P.; Hannu, J.; Norrby, T.; Byheden, Å. Additive influence on wear and friction performance of environmentally adapted lubricants. *Tribol. Int.* **2001**, *34*, 547–556. [CrossRef]
5. Li, H.; Zhang, Y.; Li, C.; Zhou, Z.; Nie, X.; Chen, Y.; Cao, H.; Liu, B.; Zhang, N.; Said, Z.; et al. Extreme pressure and antiwear additives for lubricant: Academic insights and perspectives. *Int. J. Adv. Manuf. Technol.* **2022**, *120*, 1–27. [CrossRef]
6. Ma, L.; Zhao, J.; Zhang, M.; Jiang, Z.; Zhou, C.; Ma, X. Study on the Tribological Behaviour of Nanolubricants during Micro Rolling of Copper Foils. *Materials* **2022**, *15*, 2600. [CrossRef]
7. Smith, A.T.; LaChance, A.M.; Zeng, S.; Liu, B.; Sun, L. Synthesis, properties, and applications of graphene oxide/reduced graphene oxide and their nanocomposites. *Nano Mater. Sci.* **2019**, *1*, 31–47. [CrossRef]
8. Mohan, V.B.; Lau, K.; Hui, D.; Bhattacharyya, D. Graphene-based materials and their composites: A review on production, applications and product limitations. *Compos. Part B Eng.* **2018**, *142*, 200–220.
9. Mu, J.; Gao, F.; Cui, G.; Wang, S.; Tang, S.; Li, Z. A comprehensive review of anticorrosive graphene-composite coatings. *Prog. Org. Coat.* **2021**, *157*, 106321.
10. Sang, M.; Shin, J.; Kim, K.; Yu, K.J. Electronic and Thermal Properties of Graphene and Recent Advances in Graphene Based Electronics Applications. *Nanomaterials* **2019**, *9*, 374.
11. Olabi, A.G.; Abdelkareem, M.A.; Wilberforce, T.; Sayed, E.T. Application of graphene in energy storage device—A review. *Renew. Sustain. Energy Rev.* **2021**, *135*, 110026. [CrossRef]
12. Nag, A.; Mitra, A.; Mukhopadhyay, S.C. Graphene and its sensor-based applications: A review. *Sens. Actuators A Phys.* **2018**, *270*, 177–194. [CrossRef]
13. Nasiłowska, B.; Bogdanowicz, Z.; Hińcza, K.; Mierczyk, Z.; Góźdź, S.; Djas, M.; Kowiorski, K.; Bombalska, A.; Kowalik, A. Graphene oxide aerosol deposition and its influence on cancer cells. Preliminary results. *Materials* **2020**, *13*, 4464. [CrossRef] [PubMed]
14. Ławkowska, K.; Pokrywczyńska, M.; Koper, K.; Kluth, L.A.; Drewa, T.; Adamowicz, J. Application of Graphene in Tissue Engineering of the Nervous System. *Int. J. Mol. Sci.* **2022**, *23*, 33.
15. Zhao, J.; Mao, J.; Li, Y.; He, Y.; Luo, J. Friction-induced nano-structural evolution of graphene as a lubrication additive. *Appl. Surf. Sci.* **2018**, *434*, 21–27. [CrossRef]
16. Guo, Y.; Zhang, S. The Tribological Properties of Multi-Layered Graphene as Additives of PAO2 Oil in Steel–Steel Contacts. *Lubricants* **2016**, *4*, 30–42.
17. Xi, H.; Liu, S. 2D nanomaterials as lubricant additive: A review. *Mater. Des.* **2017**, *135*, 319–332.

18. Lin, J.; Wang, L.; Chen, G. Modification of Graphene Platelets and their Tribological Properties as a Lubricant Additive. *Tribol. Lett.* **2011**, *41*, 209–215.
19. Mura, A.; Cura, F.; Adamo, F. Evaluation of graphene grease compound as lubricant for spline couplings. *Tribol. Int.* **2018**, *117*, 162–167. [CrossRef]
20. Fu, H.; Yan, G.; Li, M.; Wang, H.; Chen, Y.; Yan, C.; Lin, C.-T.; Jiang, N.; Yu, J. Graphene as a nanofiller for enhancing the tribological properties and thermal conductivity of base grease. *RSC Adv.* **2019**, *9*, 42481–42488.
21. Kinoshita, H.; Nishina, Y.; Alias, A.A.; Fujii, M. Tribological properties of monolayer graphene oxide sheets as water-based lubricant additives. *Carbon* **2014**, *66*, 720–723. [CrossRef]
22. Wang, L.F.; Ma, T.B.; Hu, Y.Z.; Wang, H. Atomic-scale friction in graphene oxide: An interfacial interaction perspective from first-principles calculations. *Phys. Rev. B* **2012**, *86*, 125436. [CrossRef]
23. Liu, Y.; Dong, Y.; Zhang, Y.; Liu, S.; Bai, Y. Effect of different preparation processes on tribological properties of graphene. *Nanomater. Nanotechnol.* **2020**, *10*, 1–8.
24. Kong, S.; Wang, J.; Hu, W.; Li, J. Effects of Thickness and Particle Size on Tribological Properties of Graphene as Lubricant Additive. *Tribol. Lett.* **2020**, *68*, 112. [CrossRef]
25. Larsson, E.; Westbroek, R.; Leckner, J.; Jacobson, S.; Rudolphi, A.K. Grease-lubricated tribological contacts—Influence of graphite, graphene oxide and reduced graphene oxide as lubricating additives in lithium complex (LiX)- and polypropylene (PP)-thickened greases. *Wear* **2021**, *486–487*, 204107. [CrossRef]
26. Chlanda, A.; Kowiorski, K.; Małek, M.; Kijeńska-Gawrońska, E.; Bil, M.; Djas, M.; Srachowski, T.; Swieszkowski, W.; Lipińska, L. Morphology and Chemical Purity of Water Suspension of Graphene Oxide Flakes Aged for 14 Months in Ambient Conditions. A Preliminary Study. *Materials* **2021**, *14*, 4108. [CrossRef]
27. Brzhezinskaya, M.; Kapitanova, O.O.; Kononenko, O.V.; Koveshnikov, S.; Korepanov, V.; Roshchupkin, D. Large-scalable graphene oxide films with resistive switching for non-volatile memory applications. *J. Alloy Compd.* **2020**, *849*, 156699. [CrossRef]
28. Brzhezinskaya, M.; Kononenko, O.V.; Matveev, V.; Zotov, A.; Khodos, I.I.; Bozhko, S.; Chekmazov, S.V.; Roshchupkin, D. Engineering of Numerous Moiré Super-lattices in Twisted Multilayer Graphene for Twistronics and Straintronics Applications. *ACS Nano* **2021**, *15*, 12358–12366. [CrossRef]
29. Gupta, V.; Sharma, N.; Singh, U.; Arif, M.; Singh, A. Higher oxidation level in graphene oxide. *Optik* **2017**, *143*, 115–124. [CrossRef]
30. Muzyka, R.; Drewniak, S.; Pustelny, T.; Chrubasik, M.; Gryglewicz, M. Characterization of Graphite Oxide and Reduced Graphene Oxide Obtained from Different Graphite Precursors and Oxidized by Different Methods Using Raman Spectroscopy. *Materials* **2018**, *11*, 1050. [CrossRef]
31. De Laurentis, N.; Kadiric, A.; Lugt, P.; Cann, P. The influence of bearing grease composition on friction in rolling/sliding concentrated contacts. *Tribol. Int.* **2016**, *94*, 624–632. [CrossRef]
32. Dhiman, C.; Reddy, M.N.; Gulati, K.; Khan, M.S. Detection of Elemental Composition of Lubricating Grease Using Laser Induced Breakdown Spectroscopy. *Lubricants* **2014**, *2*, 223–236. [CrossRef]
33. Hodapp, A.; Conrad, A.; Hochstein, B.; Jacob, K.-H.; Willenbacher, N. Effect of Base Oil and Thickener on Texture and Flow of Lubricating Greases: Insights from Bulk Rheometry, Optical Microrheology and Electron Microscopy. *Lubricants* **2022**, *10*, 55. [CrossRef]
34. Kim, C.B.; Lee, J.; Cho, J.; Goh, M. Thermal conductivity enhancement of reduced graphene oxide via chemical defect healing for efficient heat dissipation. *Carbon* **2018**, *139*, 386–392. [CrossRef]
35. Song, J.; Zhang, Y. Vertically aligned silicon carbide nanowires/reduced graphene oxide networks for enhancing the thermal conductivity of silicone rubber composites. *Compos. Part A Appl. Sci. Manuf.* **2020**, *133*, 105873. [CrossRef]
36. Gupta, B.; Kumar, N.; Panda, K.; Kanan, V.; Joshi, S.; Visoly-Fisher, I. Role of oxygen functional groups in reduced graphene oxide for lubrication. *Sci. Rep.* **2017**, *7*, 45030. [CrossRef] [PubMed]
37. Chen, W.; Yan, L. Preparation of graphene by a low-temperature thermal reduction at atmosphere pressure. *Nanoscale* **2010**, *2*, 559–563. [CrossRef] [PubMed]
38. Emiru, T.F.; Ayele, D.W. Controlled synthesis, characterization and reduction of graphene oxide: A convenient method for large scale production. *Egypt. J. Basic Appl. Sci.* **2017**, *4*, 74–79. [CrossRef]
39. Qiu, Y.; Guo, F.; Hurt, R.; Kulaots, I. Explosive thermal reduction of graphene oxide-based materials: Mechanism and safety implications. *Carbon* **2014**, *72*, 215–223. [CrossRef]

materials

MDPI

Article

Influence of Fabrication Method and Surface Modification of Alumina Ceramic on the Microstructure and Mechanical Properties of Ceramic–Elastomer Interpenetrating Phase Composites (IPCs)

Paulina Kozera [1,*], Anna Boczkowska [1], Krzysztof Perkowski [2], Marcin Małek [3] and Janusz Kluczyński [4]

1 Faculty of Materials Science and Engineering, Warsaw University of Technology, Woloska 141, 02-507 Warszawa, Poland
2 Łukasiewicz Research Network, Institute of Ceramics and Building Materials, Ceramics and Concrete Center, Postępu 9, 02-676 Warsaw, Poland
3 Faculty of Civil Engineering and Geodesy, University of Technology, Gen. Sylwestra Kaliskiego 2, 00-908 Warsaw, Poland
4 Faculty of Mechanical Engineering, Military University of Technology, ul. Gen. Sylwestra Kaliskiego 2, 00-908 Warsaw, Poland
* Correspondence: paulina.kozera@pw.edu.pl

Citation: Kozera, P.; Boczkowska, A.; Perkowski, K.; Małek, M.; Kluczyński, J. Influence of Fabrication Method and Surface Modification of Alumina Ceramic on the Microstructure and Mechanical Properties of Ceramic–Elastomer Interpenetrating Phase Composites (IPCs). *Materials* **2022**, *15*, 7824. https://doi.org/10.3390/ma15217824

Academic Editor: Alberto Ortona

Received: 11 October 2022
Accepted: 3 November 2022
Published: 6 November 2022

Abstract: The paper presents experimental results of the work conducted to improve the adhesion between alumina ceramics and urea-urethane elastomer in the interpenetrating phase composites (IPCs), in which these two phases are interpenetrating three-dimensionally and topologically throughout the microstructure. Measurements of the contact angle, surface roughness, and shear tests were used to evaluate the effectivity and select the quantity of a silane coupling agent and the ceramic fabrication method. The tests were conducted using samples of dense alumina ceramic obtained by three- or four-step methods. In the four-step process, hot isostatic pressing (HIP) was applied additionally. As a result of the coupling agent coat and HIP application, the ceramic substrate wettability by the elastomer was improved. The water contact angle was reduced from 80 to 60%. In the next step, porous ceramic preforms were fabricated using HIP sintering and a solution of silane coupling agent treated their surface. The composites were produced using vacuum-pressure infiltration of porous alumina ceramics by urea-urethane elastomer in liquid form. The influence of the coupling agent application on the microstructure and mechanical properties of the composites was estimated. The microstructure of the composites was identified using SEM microscopy and X-ray tomography. As a result of using the coupling agent, residual porosity decreased from 7 to 2%, and compressive strength, as well as stress at a plateau, increased by more than 20%, from 25 to 33 MPa and from 15 to 24 MPa, respectively, for the composites fabricated by infiltration ceramic preforms with 40% of porosity.

Keywords: IPCs; ceramic preform; ceramic–elastomer composite; silane coupling agent; hot isostatic pressing (HIP); wettability; mechanical properties

1. Introduction

Composite materials are a widely used group of materials in various industries. Traditional composites can be classified according to the shape and geometry of the reinforcement (such as particles, long fibers, short fibers, etc.). Interpenetrating phase composites (IPCs) have become a new group of materials that can be used as light materials with improved physicochemical, mechanical, and thermal properties. The IPCs are characterized by the microstructure with two three-dimensional and topologically interpenetrating phases. Furthermore, the absence of the preferred orientation of matrix and reinforcement in IPCs makes them predominantly isotropic and with consequent properties that are unattainable by other materials [1,2].

The strength of the phase separation surface, called the interphase boundary, depends on the type of phase joint, the appropriate wettability of the solid material by the liquid phase, and the surface topography of the solid component. Regardless of the type of reinforcement, the adhesion between inorganic and organic components in composites is usually weak due to the poor compatibility of the polymer or metal with the mineral surface. The problem of the insufficient wetting of the ceramics by molten metal or polymer is one of surface tension and surface quality, including any contamination, or oxidation. Some basic ways can be used to improve wetting. Generally, increasing the surface energies of the solid, decreasing the surface tension of the liquid matrix, as well as decreasing the solid–liquid interfacial energy at the interface can be essential [3]. Several treatments could be performed to increase the wettability of IPCs. In the work [4,5] it was found that the pressure-assisted infiltration, as well as process temperature and gas atmosphere, overcomes insufficient wettability between liquid metal and ceramic preform compared to pressureless fabrication methods. Moreover, in the case of pressureless techniques, insufficient wettability of the ceramic preform by liquid metal causes the formation of intermetallics as a result of interfacial reactions. According to [6,7], the application of some additives to aluminum alloys used in pressureless infiltration to the fabrication of ceramic–metal interpenetrating phase composites improved the wettability between the metallic and ceramic phases. Wang et al. [6] studied the microstructure of interface in 3D-SiC/Al-Si-Mg interpenetrating composites. It was shown that to improve wettability; the optimum content of Mg addition to the Al alloy was 4–8 wt%. A similar effect was obtained for SiC/aluminum alloy composites fabricated by pressureless infiltration as a result of the addition of 6–12 wt% Si [7]. A new method of producing ceramic-metal IPCs is the focus of the work reported by Qi et al. [8]. In the ultrasonic infiltration, the generated ultrasonic waves cause a collapse of gases dissolved in the melt and those entrapped in the ceramic foam. As a result, a pressure wave provides filling of the porous ceramic preform by the molten alloy. Furthermore, it was observed that the contact angle ZrB_2–SiC porous ceramic by aluminum alloy decreased, improving wettability.

An alternative possible technique to improve adhesion between composite phases is to modify the filler surface with a surfactant, a coupling agent, or other surface-active agents. They reduce surface tension at the interface, which has a positive effect on the adhesion between composite components [9,10]. In the literature, the application of an organosilicon coupling agent, especially for composite materials based on ceramics and polymers or metal and polymer, is widely described. When an organofunctional coupling agent is added to ceramic–polymer composites during the polymerization or crosslinking process, it forms a phase at the boundary by reacting with the functional groups of the polymer [10–12]. The use of coupling agents can increase the degree of ceramic pore filling by the polymer in a liquid form up to 80 vol.%. In addition, improving the adhesion between the components of the composite increases its compressive and tensile strength. Moreover, it can also lead to increased abrasion and corrosive resistance of the composite [11,13,14].

In polymer ceramics composites, microcracks may induce fracture to occur during stressing at or near the components' boundary. Therefore, the endurance of the connection between two phases in a composite is one of the main factors determining the IPCs' properties. To improve the mechanical properties of inorganic–organic phases, coupling agents can be applied. In the work [15], it was found that using silane coupling agents may cause the formation of siloxane bridges at the interphase, and as a consequence, result in the improvement of adhesive strength of polymer to ceramics.

In the work of [16], the effect on the dielectric and mechanical properties of co-continuous barium strontium titanate (BST)–polymer composites of surfactant introduction onto the ceramic filler surface was investigated. The result means that carboxylic acid with longer carbon chains enhances the relative permittivity of the composite more effectively. The feasibility of processing polymer syntactic–aluminum foam IPC and its compression response was studied by Jhaver and Tippur [17,18]. They applied silane to increase adhesion between the metal scaffold and polymer foam. The IPC foam coated by silane showed improvement in elastic modulus, compression strength, and plateau stress values by 28–35%, 20–25%, and 37–42% respectively.

The adhesion of two phases in IPC composites depends on the development of the ceramic surface. The topography of the ceramics' surface is changed according to the method and parameters of fabrication. Generally, during the sintering of a powder compact, both densifications occur simultaneously. The most effective way to stop grain growth and, as a result, obtain a ceramic fine-grain structure is to support the sintering processes with high pressure. Despite the high cost of the Hot Isostatic Pressing (HIP) technique, most ceramics, sintered under isostatic pressure gas, compared to freely sintered materials, are characterized by better fracture toughness [19], increased compressive strength, and tearing [20,21], as well as greater Vickers hardness [22]. Taking into consideration the porous ceramic fabrication, the HIP technique is a relatively rarely used method. Polymeric sponge method or polymer replica technique, gel casting of foams of ceramic foams all well as porous structure printing are more commonly used methods. Regardless of the method used and the pore size achieved, fabrication of a preforms structure of open pores and joined canals is required in the case of IPC materials. This structure allows for the easy flow of the liquid material.

In work [23], the authors present a comprehensive description of the freeze casting method, also known as ice templating, which has been applied extensively for the fabrication of well-controlled biomimetic porous materials based on ceramics, as well as metals. Infiltration of freeze-cast skeletons with a secondary phase allows for fabrication of aligned composites, hydrogels, and nacre-mimetic hybrids, as well as materials resembling interpenetrating phase composites.

This paper presents the influence of the fabrication methods of ceramic and its surface modification by the coupling agent application on the microstructure and mechanical properties of the ceramic–elastomer composite. Tests were carried out to obtain an accurate ceramic–elastomer joint characterization, measurements of a contact angle, surface roughness, and shear strength. The appropriate amount of coupling agent was determined for ceramic–elastomer composite fabrication. The application of a silane coupling agent improved the properties of the composites, as well as a ceramic–elastomer joint.

2. Materials and Methods

2.1. Materials

Two types of alumina ceramic samples were fabricated. The first type, dense ceramic specimens, investigated how the hot isostatic pressure sintering application affects ceramic surface roughness. Then, the dense ceramic specimens were treated with the silane coupling agent and measurements of the contact angle, and shear tests of adhesive joints were carried out. In the second type of sample, porous ceramic preforms were applied to ceramic–elastomer composites fabrication and then was analyzed how the coupling agent affected their microstructure, residual porosity, and compressive strength as well as stress at the plateau. Moreover, porous ceramic preforms were tested using X-ray tomography and a strength testing machine. All the material fabrication processes carried out are shown schematically in Figure 1.

The dense alumina ceramic samples were prepared by the three-step or four-step methods. First, the pre-compaction process of Granulox NM9922 (Nabaltec AG, Schwandorf, Germany) alumina granulate in steel molds by uniaxial pressing at a working pressure of 200 MPa was conducted. Secondly, the samples were densified into green bodies by cold isostatic pressing (CIP) under a pressure of 200 MPa for 1 min. Next, the free sintering process (pressureless sintering) was carried out in a Nabertherm HT 08/18 electric chamber furnace (Nabertherm, Lilienthal, Germany). An air inert atmosphere was used. After reaching 1600 °C, the samples were heated for 2 h to get rid of organic binders contained in the pressed shape. The samples produced by the three-step method were designated

Al_2O_3-SS. A part of the samples marked Al_2O_3-HIP200 was additionally treated by HIP process under a pressure of 200 MPa in an argon atmosphere at 1600 °C for 1 h.

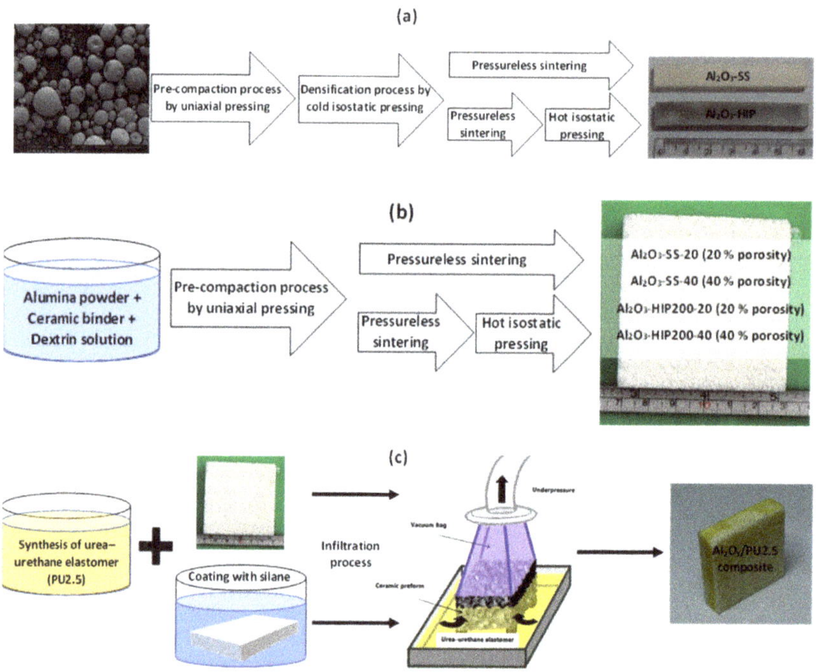

Figure 1. Scheme of materials fabrication processes: (**a**) dense ceramics fabrication, (**b**) porous ceramics preform production, (**c**) composite fabrication, respectively.

The urea–urethane elastomers (PU2.5) were synthesized by a one-shot method from 4,4′-methylenebis(phenylisocyanate) (MDI), poly(ethylene adipate) PEA, and dicyandiamide (DCDA). The molar ratio of MDI to (PEA + DCDA) substrates was equal to 2.5 (which means a hard to soft segments ratio equal to 1.50).

The porous ceramic preforms applied for the infiltration process were fabricated using the same methods as dense ceramic samples. However, to obtain open porous preforms, the alumina powders supplied by the P.P.U.H.KOS company were mixed with 10–15 wt% of Granulox NM9922 (Nabaltec AG, Schwandorf, Germany) high-temperature ceramic binder and 7 wt% of dextrin solution as pore structure forming agent. Two different sizes of alumina powders were used to fabricate ceramic preforms with different values of porosity. As a result of an application of alumina powder with 300–100 μm and 1200–1000 μm size, preforms with 20% and 40% porosities were produced, respectively. After the preparation of two types of ceramic mixtures, they were inserted into steel forms and molded by uniaxial pressing at a working pressure of 100 MPa. In the next step, the pre-sintering of semi-finished preforms was performed using an electric chamber furnace. Finally, to enhance mechanical properties and increase the density of ceramic preforms, hot isostatic pressing was conducted under 200 MPa pressure in argon at 1600 °C temperature for 1 h.

The ceramic–elastomer composites were made by the infiltration of the ceramic preforms with 20 vol% and 40 vol% porosity by a reactive mixture of urea–urethane elastomers in the liquid form. The infiltration was carried out under subatmospheric pressure.

The coupling agents can be used in two ways. In general, the surface of ceramics before infiltration by a polymer is covered by the solution of the coupling agents [11,14]. It was claimed that the most satisfying results were achieved when the agents were dissolved in benzene, toluene, and alcohol in an amount of not more than 10% by weight. In the case of ceramic–polymer composites, to cover the pore surfaces thoroughly, it is recommended to apply pressure pre-infiltration of the ceramic preforms by a coupling agent. Then, the process of drying the preform is used. This type of sample preparation is characterized by high durability and resistance during storage. The disadvantage of this method is the need to introduce into the fabrication process an additional step consisting of coating the ceramic preform with a coupling agent. Another way of applying the coupling agent is by adding it to the reactive mixture of polymer before filling the ceramic pores. During infiltration, the coupling agent migrates to the phase boundaries, acting as an adhesion promoter. The advantage of this method is the ease of implementation. However, there are several problems regarding the stability of the agent and the durability of its conjunction with the polymer, due to the presence of moisture in the system. Moreover, it was observed that a higher amount of coupling agent should be used, because a significant part of it does not reach the interface and remains in the "mass" of the polymer [10].

In this work, a Unisilan U-15 silane coupling agent (UNISIL Company, Tarnów, Poland) was applied. The U-15 Unisilan with the chemical name N-2-aminoethyl-3aminopropyl-trimethoxysilane is mainly used as a coupling agent to inorganic fillers in composites based on acrylic polymer, epoxy resin, vinyl polymer, and polyamide, polyether, polyurethane, silicone. A U-15 silane coupling agent was applied directly to the ceramic surface. Dense ceramic samples, as well as porous ceramic preforms, were immersed in the solution of 1 wt% and 5 wt% U-15 agent in toluene for 5 s. Next, the samples were placed in an oven for 2 h at 120 ± 5 °C under a vacuum to evaporate the toluene. Next, the temperature was reduced to about 80 °C and the samples were heated for another 18 h.

2.2. Methods

The roughness tests were performed using a non-contact 3D surface profiler Slynx Sensofar (Sensofar Metrology, Terrassa, Spain). Two parameters were determined: the R_a parameter, i.e., the arithmetic mean height of the surface, and the R_z parameter, i.e., the height of the highest point on the surface. The final values are the average of three different measuring points on the surfaces.

The wettability of the ceramic surfaces was tested by measuring the water contact angles (WCA) and the contact angles hysteresis (CAH) using an OCA15 (DataPhysics Instruments, Germany) goniometer equipped with OCA software. The contact angle hysteresis was determined by calculating the difference between the advancing and receding contact angles. The test was performed on rectangular tiles with dimensions of 50 × 40 × 10 mm using a 5 µL droplet. The final WCA values are the average of three different measuring points on the surfaces.

To examine the strength of the adhesive ceramic–elastomer joints, a shear test was conducted. The Lloyd LR 10K (AMETEK, Berwyn, IL, USA) strength testing machine connected with the Nexygen 3.0 computer program (AMETEK, Berwyn, IL, USA) was used for the tests. Shear strength was calculated from the force registered by the computer divided by the joint cross-section according to the formula (1). The tests were carried out at a constant velocity of 1 mm/min according to the ASTM D3163-01 standard:

$$\tau_{max} = \frac{F_{max}}{A} \text{ (MPa)} \tag{1}$$

where: τ_{max} is the maximum shear stress [MPa], F_{max} is the force causing damage to the joint (N), and the A is the cross-sectional area of the joint (mm^2).

Adhesive joints were made for the test according to the scheme shown in Figure 2. Two ceramic plates were placed in a Teflon mold. A layer of elastomer was placed between them. The thickness of the elastomer layer corresponded to the thickness of the spacer plate. The ceramic specimens with dimensions of 5 × 7 × 37 mm were first placed in a molder, and then the reactive mixture was poured.

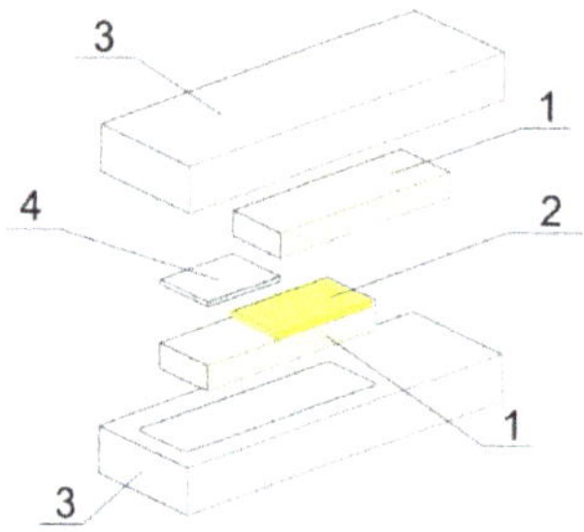

Figure 2. Graph showing the construction of the adhesive joint, 1—dense ceramic substrate, 2—elastomer, 3—mold, 4—spacer.

The character of joint failure (adhesive or cohesive) and the microstructure of composites were characterized using Scanning Electron Microscopy (SEM) TM3000 (HITACHI High-Technologies Corporation, Tokyo, Japan) operating at an applied voltage of 5 kV. Before observations with the SEM, the surfaces of the specimens were sputtered with a gold-palladium layer for 90 s at a current of 10 mA and voltage of 2 kV. SkyScan 1174 X-ray tomography (SkyScan, Aartselaar, Belgium) was used for testing of the ceramic preform and composites. Before scanning, samples in the shape of a cuboid with dimensions $10 \times 10 \times 15$ mm did not require any special preparation. Scanning was performed using an X-ray tube with the following parameters: 100 kV voltage, 100 kA, no filter material, 0.5° rotation step in an angle interval of 180°. The obtained cross-sections of the ceramic preforms and composites were studied using CTAn software v.1.18 (Bruker, Kontich, Belgium) and as a result, porosity of the ceramic preform as well as the residual porosity of composites were determined. The application of CTAn software enabled the investigation of the weight fraction of each phase, including voids in ceramic preforms as ceramics porosity and in composites as residual porosity.

The compressive test was carried out using the MTS Q/Test 10 testing machine test machine (MTS Testing Systems, Toronto, ON, Canada) according to the ISO 20504:2019 standard with 1 mm/min velocity. Based on the obtained stress–strain curves, compressive strength and stress at the plateau were calculated.

3. Results and Discussion

3.1. *Surface Roughness Test of Dense Alumina Ceramics*

The topography of the ceramic surface was analyzed by the results of the roughness test. Results of Ra and Rz roughness parameters, as well as a surface profile observation, are displayed in Figures 3–5. The surface roughness test was conducted for dense Al_2O_3 ceramics fabricated by two methods; for Al_2O_3-SS samples, pressureless sintering was the final process, whereas for Al_2O_3-HIP200 samples hot isostatic pressure sintering was applied additionally.

The roughness parameters determined with the use of a laser profilometer allowed the assessment of the degree of surface development of both types of ceramic materials. It was demonstrated that the use of hot isostatic pressure during the sintering of alumina ceramics modified their surface condition. It should be noted that the application of 200 MPa pressure during ceramic sintering caused a decrease in roughness parameters in comparison to the pressureless sintered material. Looking into the ceramic profiles, in the Al_2O_3-HIP200 sample, numerous structural defects with a size of about 50 µm and a depth of 9 µm were eliminated. Other researchers have previously observed a similar relationship in alumina ceramic sintering [24]. In comparison, Al-Jawoosh et al. indicated that the R_a roughness parameter for the densely sintered alumina ceramic specimens was 0.6 µm [25]. As a consequence of the use of high pressure, the pores and cracks closing occurred and the material's compaction increased. Furthermore, densification was due to particle rearrangement, plastic deformation, grain boundary diffusion, and structural defect elimination. During the sintering of a powder compact, both densification and grain growth occur simultaneously [26].

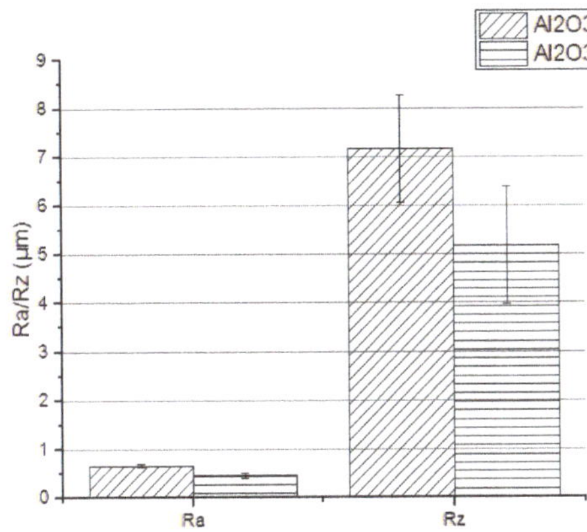

Figure 3. The R_a and R_z roughness parameters of dense alumina ceramics depend on fabrication methods: Al_2O_3-SS samples fabricated by pressureless sintering and Al_2O_3-HIP200 specimens obtained using hot isostatic pressure sintering.

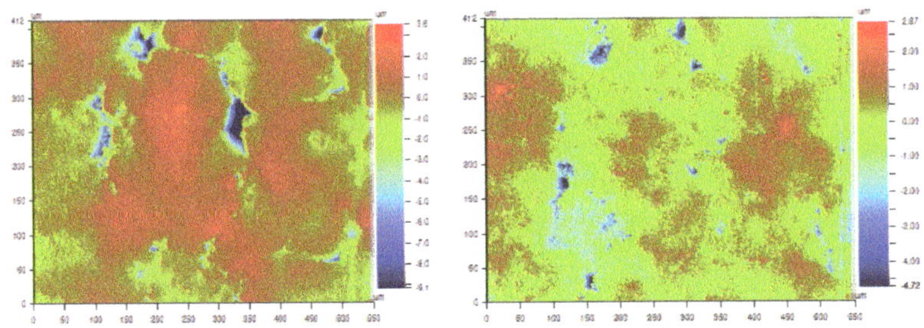

(**a**) (**b**)

Figure 4. The surface topography map of dense alumina ceramics depends on fabrication methods: (**a**) Al_2O_3-SS samples fabricated by pressureless sintering and (**b**) Al_2O_3-HIP200 specimens obtained using hot isostatic pressure sintering.

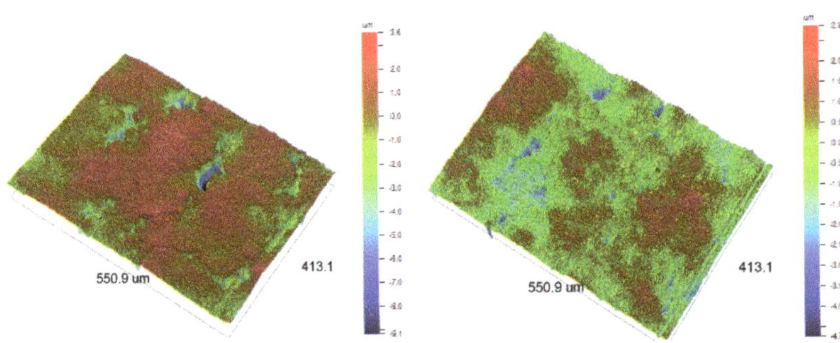

(**a**) (**b**)

Figure 5. 3D Profilometry images illustrating dense alumina ceramics depending on fabrication methods: (**a**) Al_2O_3-SS samples fabricated by pressureless sintering and (**b**) Al_2O_3-HIP200 specimens obtained using hot isostatic pressure sintering.

3.2. The Contact Angle Measurements of Dense Alumina Ceramics

The contact angle measurements were performed to examine the effect of the HIP sintering application as well as the silane coupling agent on ceramic wettability. Furthermore, the influence of the coupling agent content in the toluene solution on the contact angle was investigated. The results of the tests are presented in Figure 6.

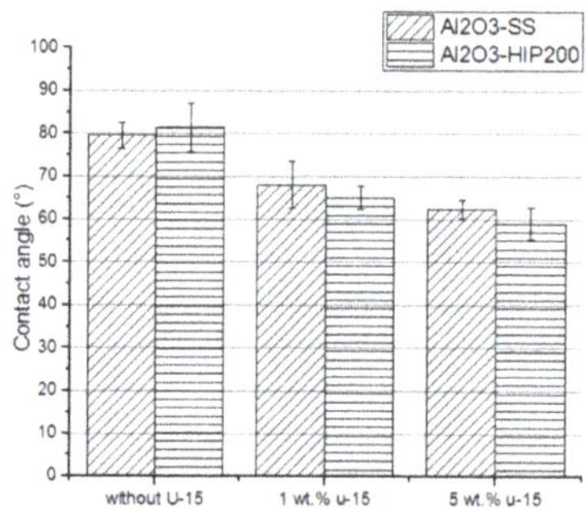

Figure 6. The contact angles on the alumina ceramic surface were treated using a silane solution of the coupling agent with different concentrations; Al_2O_3-SS samples were fabricated by pressureless sintering, and Al_2O_3-HIP200 specimens were obtained using HIP sintering.

Wettability is the ability to spread material in a liquid form on a solid surface. Depending on the contact angle value, surfaces are of a hydrophilic character ($\theta < 90°$) or hydrophobic character ($\theta > 90°$). In the case of IPC composites, the value of the contact angle affects the degree of filling of the porous preform by the liquid material, and thus affects the mechanical strength of the composite [27]. Reducing the contact angle value, i.e., improving wettability, can ensure better filling of pores in the ceramic preform, and elimination of gas bubbles from the interface. In addition, the elimination of structural discontinuities increases the contact area between the materials, which in turn leads to an improvement of the durability of the interface connection [28].

The obtained results show that the average water contact angle of the nontreated ceramic surface was about 80° (lack of differences between the fabrication methods of ceramic) but in the presence of a solution coat of 5 wt% U-15 coupling agent in toluene, it was decreased to 60°. This means that the wettability of ceramic was improved due to the U-15 agent application. Although an angle value below 90° indicates the hydrophilic character of the ceramic surface uncoated by the solution of the U-15 agent, the contact angle was still close to 90°. It decreased significantly only after applying the coupling agent coat.

The solution concentration of the silane coupling agent in toluene has been selected based on previous work [29]. For low concentrations of the U-15 agent (<1%), a change in water contact angle was not observed. These observations were confirmed by other researchers [16,28]. It was found that the solution concentration, solution pH, and curing conditions can affect significantly the silane bonding to an inorganic surface. Silane solution baths are usually used in low concentrations (0.01–2%) because the formation of oligomers is suppressed in dilute solutions [28].

The literature indicates that the contact angle is closely related to surface roughness [30]. It can be concluded that if the R_a parameter possesses a value lower than 0.5, the effect of roughness on the contact angle is insignificant. Higher R_a values indicate an increase in surface roughness, hence, an increase in the wetting surface (contact surface). Nevertheless, increasing roughness can affect wettability in two ways. Together with the improvement in surface roughness, the number of defects and pores on the surface into which liquid material can penetrate increases. As a result, the strength of such a joint can enhance. However, too many pores, especially narrow ones, become an obstacle, impede wettability, and consequently, prevent the formation of a durable adhesive interface connection [30]. As shown in Figure 2, the achieved Ra roughness parameter is low independently of the fabrication method of dense alumina ceramic. This confirms the lack of a significant difference in water contact angle results for Al_2O_3-SS and Al_2O_3-HIP200 ceramic.

3.3. *Shear Strength of an Interface Joint*

The shear strength results of the ceramic–elastomer interface joint are displayed in Figure 7. It can be seen that the application of the silane solution coat enhanced the shear strength of the ceramic–elastomer joints. The average shear strength value for Al_2O_3-SS ceramic–elastomer joint was 2.9 ± 0.7 MPa while using a silene solution coat increased the shear strength to 8.8 ± 0.2 MPa. In comparison, Chaijareenont et al. indicated that silane coupling agents affect polymethyl methacrylate (PMMA) bonding to alumina. The bond strength of PMMA on the alumina treated by a bath in a solution of N-2 (aminoethyl) 3-aminopropyltriethoxysilane) in ethanol was reached at 10.8 MPa, which is similar to obtained results [31]. The highest shear strength was achieved for the ceramic–elastomer joint in which dense ceramic was a fabrication by pressureless sintering. This is related to the slightly higher roughness parameters of the surface specimen. In other words, a higher surface roughness caused an increase in the contact area of the joint materials, while the use of a coupling agent ensured the possibility to infiltrate the micropores on the ceramic surface by the elastomer reactive mixture. It is noteworthy that, in the case of dense ceramic fabricated by pressureless sintering, the pore size was not reduced as for Al_2O_3-200HIP samples. Hence, the penetration of the reactive mixture into the micropores of the ceramic surface and the creation of additional mechanical bonds was easier. In addition, due to its bi-functionality, the U-15 agent interacted with two materials, improving their mutual adhesion by creating molecular bridges, which connected the inorganic surface with the polymer through the available types of polar interaction. The bonding was covalent (siloxane bond) [32].

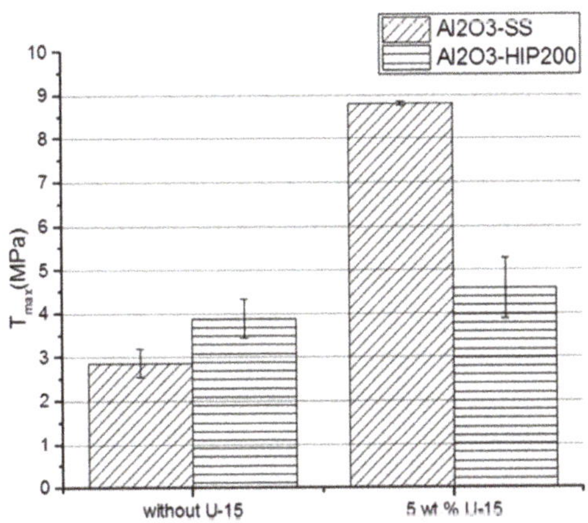

Figure 7. The shear strength on the alumina ceramic surface was treated using a silane solution of the coupling agent with different concentrations; Al_2O_3-SS samples were fabricated by pressureless sintering, and Al_2O_3-HIP200 specimens were obtained using HIP sintering.

3.4. *SEM Observations of Ceramic–Elastomer Joint*

The ceramic–elastomer bond quality was analyzed in terms of failure character joint after a shear test using a scanning electron microscope. Sample images are presented in Figure 8a,b. At first glance, it can be seen that the failure was adhesive for interfacial joints in which no coupling agent was applied, as evidenced by the smooth surface of the joint after the shear test. In contrast, in samples in which the coupling agent was applied

(Figure 8b), the failure was reached partially by elastomer decohesion, which indicated good adhesion of the elastomer to the ceramic surface.

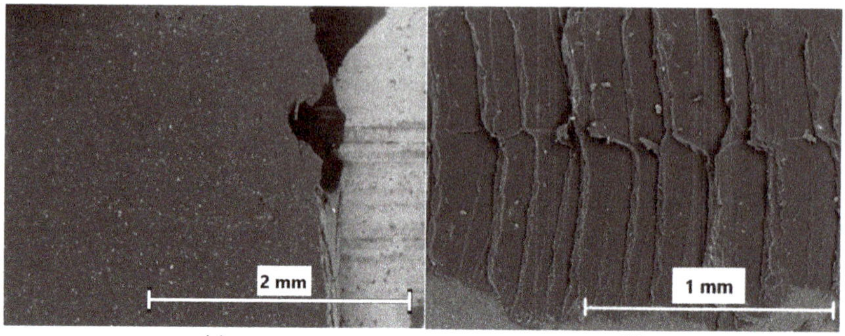

(**a**) (**b**)

Figure 8. SEM images illustrating failure after shear strength test of ceramic–elastomer joints: (**a**) alumina surface uncoated of silane solution, (**b**) alumina surface coated of silane solution.

3.5. Mechanical Properties of Porous Alumina Ceramic

The main property of porous ceramic preforms is high mechanical strength. Therefore, the effect of the fabrication method as well as the degree of porosity of ceramic preform into their compressive strength were analyzed (Table 1). The results show that the average compressive strength of the sample fabricated by pressureless sintering was about 10 MPa, but with the addition of the HIP sintering stage, it was increased by 100%. This is related to the densification of alumina grains and pores size reduction.

Table 1. Compressive strength of porous ceramic preform fabricated by pressureless and HIP sintering.

Material	Type of Applied Final Stage of a Fabrication Method	Porosity [%]	Compressive Strength [MPa]
Al_2O_3-SS-20 Al_2O_3-SS-40	Pressureless sintering	22 ± 0.5 41 ± 1.0	10.2 ± 2.1 6.5 ± 1.2
Al_2O_3-HIP200-20 Al_2O_3-HIP200-40	Hot isostatic pressing	23 ± 0.7 40.5 ± 2.0	21.3 ± 3.0 17.6 ± 1.4

Ceramic preforms intended for pressure infiltration should exhibit mechanical strength. Otherwise, they may be damaged during the fabrication of composites. Looking into the compressive test result, it can be seen that porous preform fabricated using HIP sintering is characterized by higher mechanical strength.

Interestingly, the degree of porosity of ceramic preform achieved in the work is considerably lower in comparison to ceramic preform porosity that can be obtained using other methods. In comparison, Peng et al. indicated that the porosity of ceramic foams fabricated by the polymeric sponge method was higher than 70% [33]. Similarly, the application of gel casting of ceramic foams method allowed porosity to be reached between 50% and 90% [34]. However, it is noteworthy that the growth in porosity causes a decrease in mechanical strength.

3.6. SEM Observations of Composites

After characterization of porous ceramic preforms' mechanical properties, a part of them was used for ceramic–elastomer composites fabrication. The mechanical test results of porous samples confirmed that the HIP sintering application allowed them to achieve higher compressive strength. From this point of view, the porous ceramic preforms were utilized for ceramic–elastomer fabrication by the infiltration method. The Al_2O_3-HIP200 porous samples with 20% and 40% porosity were infiltrated by the liquid elastomer. Part of the samples was coated with 5 wt% silane solution in toluene. To assess the microstructure of the fabricated ceramic–elastomer composites and the effect of the coupling agent on their microstructure, SEM observations were conducted. The obtained results are shown in Figures 9b and 10b. In contrast, observations of composite fabricated by using porous ceramic preform without a U-15 agent solution coat are displayed in Figures 9a and 10a.

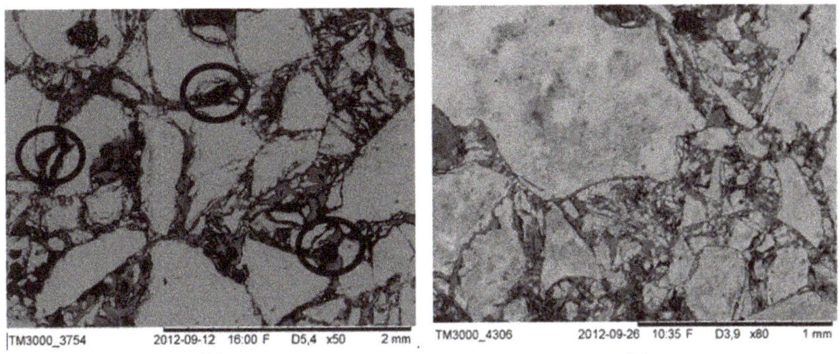

(a) (b)

Figure 9. The microstructure of Al_2O_3/PU2.5 composite fabricated by infiltration of the ceramic preform with 20% porosity: (**a**) porous alumina surface uncoated of silane solution, (**b**) porous alumina surface coated of silane solution. (The lighter phase is alumina ceramics and the darker phase is polyurethane elastomer, the black circle show structural defects).

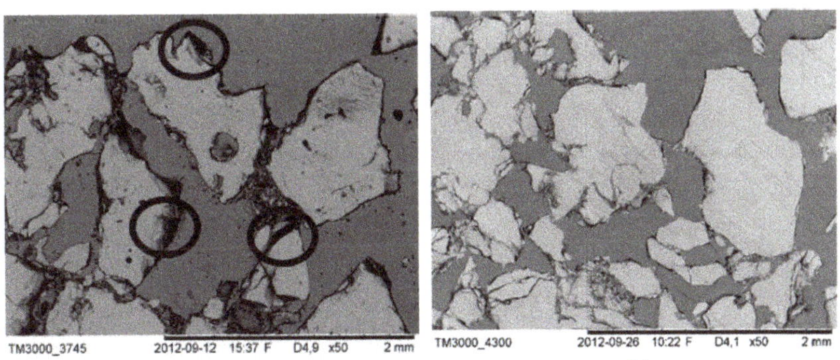

(a) (b)

Figure 10. The microstructure of Al_2O_3/PU2.5 composite fabricated by infiltration of the ceramic preform with 40% porosity: (**a**) porous alumina surface uncoated of silane solution, (**b**) porous alumina surface coated of silane solution. (The lighter phase is alumina ceramics and the darker phase is polyurethane elastomer, the black circle show structural defects).

At first glance, it can be seen that the macro and micro-pores of the ceramic preforms have been filled by the reactive mixture of elastomer. The elastomer also infiltrated the channels formed between the ceramic grains, as well as in their cracks. It should be also noticed that the structure of interpenetrating phases was successfully obtained. Observations confirmed that the infiltration method allowed the elastomer to fill the ceramic pores. Looking into the SEM images, it can be seen that the adhesion between Al_2O_3 ceramics with the elastomer for each porosity was improved for composites fabricated by using porous ceramic preform with a U-15 agent solution coat. Furthermore, the pores have been filled better and the interface boundary between the ceramics and the elastomer was continuous. In addition, delamination on the ceramic–elastomer boundary was not observed, contrary to composites obtained using porous ceramic preform without a U-15 agent solution coat. In the case of uncoated samples, the effect was revealed of weaker adhesion between the ceramic and polymer phases, as evident from the isolated debonds highlighted in Figures 9a and 10a. It can be concluded that the silane coupling agent facilitates infiltration and improves adhesion between the phases; other researchers have previously observed a similar relationship by coupling agent application [35–37].

It is noteworthy that after the ceramic pores are filled with elastomer in liquid form, polymerization of the monomer is started, and the unavoidable volume shrinkage can appear. In work [10], it was confirmed that the application of pressure during infiltration as well as a reduced polymerization speed can limit the occurring polymerization shrinkage, which leads to a decrease in the appearance of defects in the microstructure of IPCs. In this study, despite pressure infiltration application to fill ceramic pores with liquid elastomer, the polymerization shrinkage generated pores (defects) in the microstructure. This led to an interfacial boundary loss between polymer and ceramic. An optimized, defect-free microstructure was obtained for composites fabricated by infiltration of ceramic preforms coated by silane agent.

3.7. Residual Porosity Measurement of Composites

Residual porosity, i.e., porosity, which is a result of insufficient pore filling by the elastomer, was determined by X-ray tomography. The results of residual porosity for composites fabricated by infiltration of porous ceramic pre-form coated with a U-15 agent solution as well as uncoated ceramic preforms are shown in Figure 11.

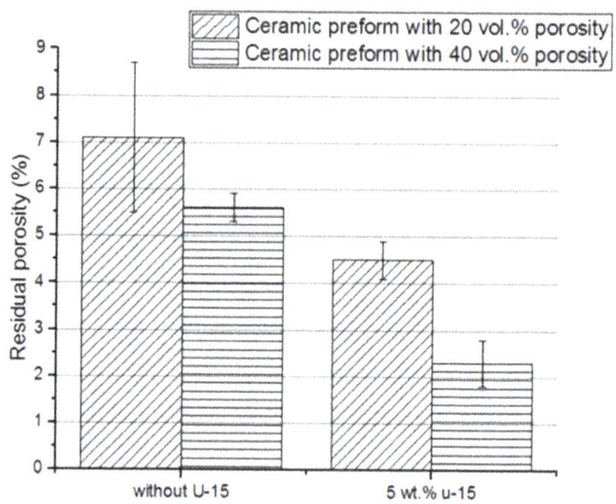

Figure 11. Comparison between residual porosity of composites fabricated by using porous alumina uncoated and coated of silane solution; Al_2O_3/PU2.5 composites fabricated by infiltration of the ceramic preform with 20% and 40% porosity.

The coupling agent had a significant impact on the infiltration process and, consequently, the degree of pore filling. The smallest residual porosity after infiltration, approximately 2 vol.%, was evaluated for composites fabricated by infiltration of ceramic pre-form with 40 vol.% porosity coated with a U-15 agent solution. Infiltration of preforms with higher porosity was easier, pores were better filled, and residual porosity was smaller than in the case of the preform with smaller porosity.

3.8. Mechanical Properties of Composites

To analyze how the coupling agent application affects the compressive strength and stress at the plateau of composites, the compression test was performed. The calculated compressive strength and stress at the plateau area are presented in Figures 12 and 13. For the composites fabricated by infiltration of the uncoated porous preform, the compressive strength was under 30 MPa. Using silane solution on porous ceramic preforms increased the compressive strength of the composite to about 35 MPa. A similar tendency of stress in the plateau area growth was found. The obtained results can be summarized: the application of the U-15 coupling agent caused a significant increase in the mechanical properties of ceramic–elastomer interpenetrating phase composites. In addition, the higher score of the area under the stress-strain curve, i.e., the plateau area, as well as stress at the plateau, was achieved. Most likely, the ability to absorb the energy of composites fabricated with a U-15 promoter application was also improved. Moreover, the samples were not destroyed during the compressive test. After removing the load, the composite specimens almost returned to their original shape, because of the highly elastic deformations of the elastomer. Similar results were obtained in [18], where the increase in elastic modulus and compressive strength of silane-coated preform can be attributed to improved wettability, which in turn enhances adhesion between the metal and polymer phases.

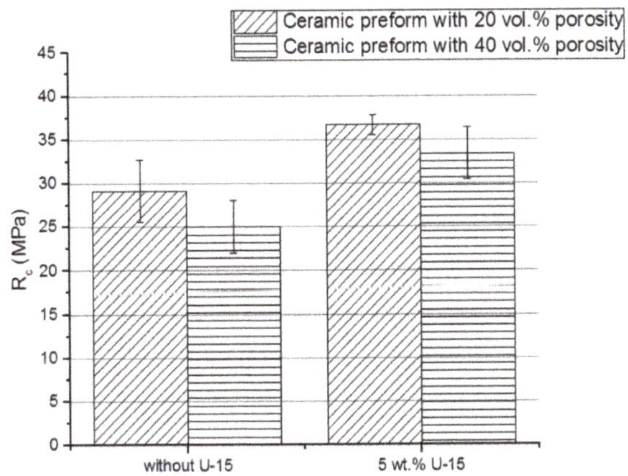

Figure 12. Comparison between compressive strength of composites fabricated by using porous alumina uncoated and coated of silane solution; Al_2O_3/PU2.5 composites fabricated by infiltration of the ceramic preform with 20% and 40% porosity.

Figure 14 shows the typical stress–strain response at static compressions of IPCs. For all composites coated and untreated with silane solution, a linear elastic deformation followed by a protracted nonlinear behavior was observed. After reaching the elastic limit, the inelastic stage includes a distinct softening response due to the onset of ceramic foam failure. It is important to note that the plateau stage is characterized by a long duration of slightly increased stress and quickly increased strain. The last stage is concerned with the densification of the composite structure due to load impact. As noted earlier, the increase in compressive strength and stress of plateau of composites fabricated by using silane-coated preform can be attributed to improved wettability, which in turn enhances adhesion between the ceramic and polymer phases. The improvement in the silane-coated composites' mechanical characteristics relative to the uncoated one is caused by the stronger bond between the ceramic and polymer matrix. Application of silane coupling agent delays failure of interfacial bonds during the deformation process [17,18].

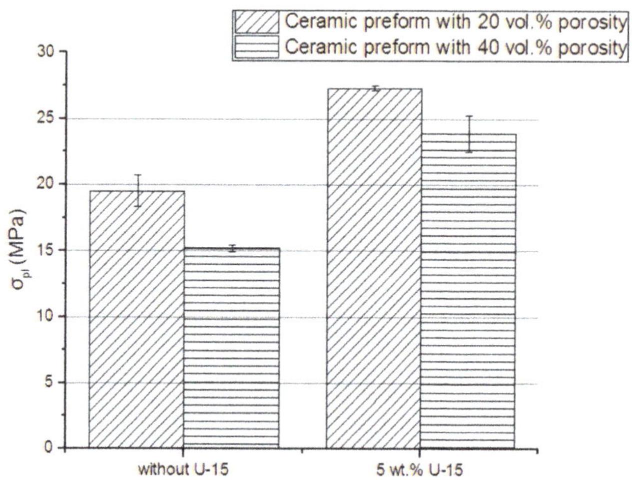

Figure 13. Comparison between stress at a plateau of composites fabricated by using porous alumina, uncoated and coated, of silane solution; Al_2O_3/PU2.5 composites fabricated by infiltration of the ceramic preform with 20% and 40% porosity.

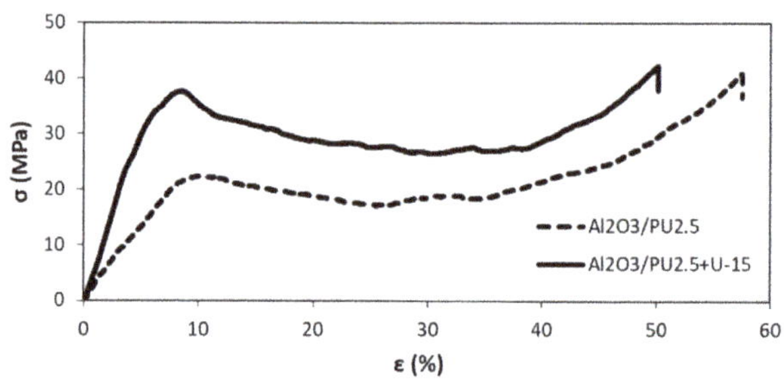

Figure 14. Stress–strain response in uniaxial compression for composites with uncoated (Al_2O_3/PU2.5) and silane coated performs (Al_2O_3/PU2.5+U-15); ceramic foam with 20% porosity.

4. Conclusions

This paper describes the influence of surface modification and fabrication methods of alumina ceramic on the adhesion between ceramic and elastomer in interpenetrating phase composites. The application of the HIP process for the fabrication of ceramic foams is rare, due to the difficulty of gaining IPCs afterward. Owing to the use of HIP, a low degree of porosity of the ceramic preforms was achieved, which provides higher mechanical properties of the composites. However, for such porosity, it is difficult to obtain a composite structure with interpenetrating of phases. Hence, silane coat applications were analyzed. The tests were carried out for dense and porous alumina ceramic as well as alumina–elastomer composites. The porous ceramic preforms were obtained by sintering methods whereas composites used infiltration of reactive elastomer mixture into ceramic pores. The fabricated composites were characterized by the microstructure with two three-dimensional and topologically interpenetrating phases. To improve the adhesion between alumina ceramics and urea-urethane elastomer in IPCs, hot isostatic pressing of alumina ceramic was utilized. Moreover, the porous ceramic preform was coated by coupling agent solution. The impact of HIP sintering and silane solution coat was investigated on roughness, wettability, microstructure, as well as mechanical properties of materials based on alumina ceramic.

The results are summarized as follows:

1. The application of hot isostatic pressure in the fabrication process of solid alumina ceramic affects slightly the roughness of the ceramic surface; however, in the case of porous ceramic, preforms' mechanical strength increases by over 100%, from 10.2 MPa to 21.3 MPa for samples with 20% porosity fabricated by HIP application.
2. Silane solution coat application achieved a significant improvement in the ceramic surface's wettability. The water contact angle was reduced from 80% to 60%.
3. A decrease in the contact angle facilitated the infiltration process of the porous ceramic preform by the elastomer. As a result, the degree of filling of the pores by the elastomer reactive mixture was enhanced for the ceramic–elastomer composite. Moreover, the residual porosity of composites decreased to 2%.
4. The mechanical properties of composites, such as compressive strength and stress at the plateau increased for composites fabricated using porous ceramic coated by the solution of coupling agent. As a result of using the silane coat, compressive strength, as well as stress at a plateau, increased by more than 20%, from 25 MPa to 33 MPa and from 15 MPa to 24 MPa, respectively, for the composites fabricated by infiltration ceramic preforms with 40% of porosity.

Author Contributions: Conceptualization, P.K.; methodology, P.K.; software, P.K.; validation, P.K., A.B., K.P., M.M., J.K.; formal analysis, P.K. and A.B.; investigation, P.K.; resources, P.K. and K.P.; data curation, P.K., A.B., K.P.; writing—original draft preparation, P.K.; writing—review and edit, P.K., A.B., K.P., M.M., J.K.; visualization, P.K.; supervision, P.K. and A.B.; project administration, P.K., M.M. and J.K.; funding acquisition, M.M. and J.K. All authors have read and agreed to the published version of the manuscript.

Funding: This research was funded by a Scholarship from the Dean of Faculty of Civil Engineering and Geodesy, Military University of Technology, number 1/DPH/2020.

Institutional Review Board Statement: Not applicable.

Informed Consent Statement: Not applicable.

Data Availability Statement: All data are included in the manuscript.

Acknowledgments: The authors would like to acknowledge Artur Oziębło and Adam Witek for support in ceramic specimens fabrication, Zbigniew Jaegermann for shear tests carrying out, Oskar Czemierowski for contact angle measurements, and Maciej Spychalski for topography investigation.

Conflicts of Interest: The authors declare no conflict of interest.

References

1. Fan, Z.; Zhang, B.; Liu, Y.; Suo, T.; Xu, P.; Zhang, J. Interpenetrating phase composite foam based on porous aluminum skeleton for high energy absorption. *Polym. Test.* **2021**, *93*, 106917. [CrossRef]
2. Aldrich, D.E.; Fan, Z. Microstructural characterisation of interpenetrating nickel/alumina composites. *Mater. Charact.* **2001**, *47*, 167–173. [CrossRef]
3. Shouren, W.; Haoran, G.; Jingchun, Z.; Yingzi, W. Interpenetrating microstructure and properties of Si_3N_4/Al–Mg composites fabricated by pressureless infiltration. *Appl. Compos. Mater.* **2006**, *13*, 115–126. [CrossRef]
4. Roy, S.; Schell, K.G.; Bucharsky, E.C.; Weidenmann, K.A.; Wanner, A.; Hoffmann, M.J. Processing and characterization of elastic and thermal expansion behaviour of interpenetrating Al12Si/alumina composites. *Mater. Sci. Eng. A* **2019**, *743*, 339–348. [CrossRef]
5. Scherm, F.; Volkl, R.; Neubrand, A.; Bosbach, F.; Glatzel, U. Mechanical characterisation of interpenetrating network metal–ceramic composites. *Mater. Sci. Eng. A* **2010**, *527*, 1260–1265. [CrossRef]
6. Wang, D.; Zheng, Z.; Lv, J.; Xu, G.; Zhou, S. Interface design in 3D-SiC/Al-Si-Mg interpenetrating composite fabricated by pressureless infiltration. *Ceram. Int.* **2018**, *44*, 11956–11965. [CrossRef]
7. Pech-Canul, M.I.; Katz, R.N.; Makhlouf, M.M.; Pickard, S. The role of silicon in wetting and pressureless infiltration of SiC preforms by aluminum alloys. *J. Mater. Sci.* **2000**, *35*, 2167–2173. [CrossRef]
8. Qi, Y.; Chen, G.; Li, Z.; Chen, L.; Han, W.; Du, Z. A novel approach to fabricate ceramic/metal interpenetrating phase composites by ultrasonic-assisted spontaneous infiltration. *Ceram. Int.* **2021**, *47*, 2903–2907. [CrossRef]
9. Boczkowska, A.; Konopka, K.; Schmidt, J.; Kurzydłowski, K.J. Investigation of elastomer structure and adhesion effect on compression strength of ceramic-elastomer composites. *Kompoz. (Compos.)* **2004**, *9*, 41–46.

10. Steier, V.F.; Koplin, C.; Kailer, A. Influence of pressure-assisted polymerization on the microstructure and strength of polymer-infiltrated ceramics. *J. Mater. Sci.* **2013**, *48*, 3239–3247. [CrossRef]
11. Zhou, R.; Lu, D.H.; Jiang, Y.H.; Li, Q.N. Mechanical properties and erosion wear resistance of polyurethane matrix composites. *Wear* **2005**, *259*, 676–683. [CrossRef]
12. Marciniec, B.; Guliński, J. Silane coupling agents. *Polimery* **1995**, *2*, 77–81. [CrossRef]
13. Szafran, M.; Rokicki, G.; Bobryk, E.; Lamenta, A. Ceramics-polymer composites based on porous ceramic material with porosity gradient. *Kompoz. (Compos.)* **2004**, *4*, 231–236.
14. Colde, A.; Swain, M.V.; Thiel, N. Mechanical properties of polymer-infiltrated-ceramic-network materials. *Dent. Mater.* **2013**, *29*, 419–426. [CrossRef] [PubMed]
15. Miyata, N.; Matsuura, W.; Kokubo, T.; Nakamura, T. Mechanical behavior of bioactive composite cements consisting of resin and glass-ceramic powder in a simulated body fluid: Effect of silane coupling agent. *J. Mater. Sci. Mater. Med.* **2004**, *15*, 1013–1020. [CrossRef]
16. Sonoda, K.; Juuti, J.; Moriya, Y.; Jantunen, H. Modification of the dielectric properties of 0-3 ceramic-polymer composites by introducing surface active agents onto the ceramic filler surface. *Compos. Struct.* **2010**, *92*, 1052–1058. [CrossRef]
17. Jhaver, R.; Tippur, H. Compression response of syntactic foam based interpenetrating phase composites. In Proceedings of the 11th International Congress and Exposition, Orlando, FL, USA, 2–5 June 2008.
18. Jhaver, R.; Tippur, H. Processing, compression response and finite element modeling of syntactic foam based interpenetrating phase composites (IPC). *Mater. Sci. Eng. A* **2009**, *499*, 507–517. [CrossRef]
19. Muchtar, A.; Lim, L.C. Indentation fracture toughness of high purity submicron alumina. *Acta Mater.* **1998**, *46*, 1683–1690. [CrossRef]
20. Rice, R.W. Review ceramic tensile strength-grain size relations: Grain sizes, slopes, and branch intersections. *J. Mater. Sci.* **1997**, *32*, 1673–1692. [CrossRef]
21. Koo, J.B.; Hong, K.J.; Park, J.S.; Shin, D.C. Effect of grain size on transmittance and mechanical strength of sintered alumina. *Mater. Sci. Eng.* **2004**, *374*, 191–195.
22. Rice, R.W.; Wu, C.C.; Borchelt, F. Hardness–Grain-size relations in ceramics. *J. Am. Ceram. Soc.* **1994**, *78*, 2539–2553. [CrossRef]
23. Shao, G.; Hanaor, D.A.; Shen, X.; Gurlo, A. Freeze Casting: From Low-Dimensional Building Blocks to Aligned Porous Structures—A Review of Novel Materials, Methods, and Applications. *Adv. Mater.* **2020**, *32*, 17. [CrossRef] [PubMed]
24. He, Z.; Ma, J. Densification and grain growth during interface reaction controlled sintering of alumina ceramics. *Ceram. Int.* **2001**, *27*, 261–264. [CrossRef]
25. Al-Jawoosh, S.; Ireland, A.; Su, B. Characterisation of mechanical and surfaceproperties of novel biomimetic interpenetratingalu mina-polycarbonate composite materials. *Dent. Mater.* **2020**, *36*, 1595–1607. [CrossRef]
26. Figiel, P.; Rozmus, M.; Smuk, B. Properties of alumina ceramics obtained by conventional and non-conventional methods for sintering ceramics. *J. Achiev. Mater. Manuf. Eng.* **2011**, *48*, 29–34.
27. Thomas, S.; Raman, S.; Mohanan, P.; Sebastian, M.T. Effect of coupling agent on the thermal and dielectric properties of PTFE/$Sm_2Si_2O_7$ composites. *Compos. A* **2010**, *41*, 1148–1155. [CrossRef]
28. Hoikkanen, M.; Honkanen, M.; Vippola, M.; Lepistö, T.; Vuorinen, J. Effect of silane treatment parameters on the silane layer formation and bonding to thermoplastic urethane. *Prog. Org. Coat.* **2011**, *72*, 716–723. [CrossRef]
29. Chabera, P.; Boczkowska, A.; Witek, A.; Oziębło, A. Fabrication and characterization of lightweight ceramic/ polyurethane composites. *Bull. Pol. Acad. Sci. Tech. Sci.* **2015**, *63*, 193–199.
30. QI, Z.; Liao, L.; Wang, R.; Zhang, Y.; Yuan, Z. Roughness-dependent wetting and surface tension of molten lead on alumina. *Trans. Nonferrous Met. Soc. China* **2021**, *31*, 2511–2521. [CrossRef]
31. Chaijareenont, P.; Takahashi, H.; Nishiyama, N.; Arksornnukit, M. Effects of silane coupling agents and solutions of different polarity on PMMA bonding to alumina. *Dent. Mater.* **2014**, *31*, 4. [CrossRef]
32. Lung, C.Y.K.; Matinlinna, J.P. Aspects of silane coupling agents and surface conditioning in dentistry: An overview. *Dent. Mater.* **2012**, *28*, 467–477. [CrossRef] [PubMed]
33. Peng, H.X.; Fan, Z.; Evans, J.R.G.; Busfield, J.J.C. Microstructure of ceramic foams. *J. Eur. Ceram. Soc.* **2000**, *20*, 807–813. [CrossRef]
34. Zeschky, J.; Goetz-Neunhoeffer, F.; Neubauer, J.; Jason Lo, S.H.; Kummer, B.; Scheffler, M.; Greil, P. Preceramic polymer derived cellular ceramics. *Compos. Sci. Technol.* **2003**, *63*, 2361–2370. [CrossRef]
35. Miller, A.C.; Berg, J.C. Effect of silane coupling agent adsorbate structure on adhesion performance with a polymeric matrix. *Compos. A* **2003**, *34*, 327–332. [CrossRef]
36. Boczkowska, A.; Konopka, K.; Kurzydłowski, K.J. Effect of elastomer structure on ceramic–elastomer composite properties. *J. Mater. Process. Technol.* **2006**, *175*, 40–44. [CrossRef]
37. Navya Kota, N.; Charan, M.S.; Laha, T.; Roy, S. Review on development of metal/ceramic interpenetating phase composites and critical analysis of their properties. *Ceram. Int.* **2022**, *48*, 1451–1483. [CrossRef]

materials

MDPI

Article

Gaussian Process for Machine Learning-Based Fatigue Life Prediction Model under Multiaxial Stress–Strain Conditions

Aleksander Karolczuk [1,*], Dariusz Skibicki [2] and Łukasz Pejkowski [2]

[1] Department of Mechanics and Machine Design, Opole University of Technology, Ul. Mikołajczyka 5, 45-271 Opole, Poland
[2] Faculty of Mechanical Engineering, UTP University of Science and Technology, Kaliskiego 7, 85-796 Bydgoszcz, Poland
* Correspondence: a.karolczuk@po.edu.pl

Abstract: In this paper, a new method for fatigue life prediction under multiaxial stress-strain conditions is developed. The method applies machine learning with the Gaussian process for regression to build a fatigue model. The fatigue failure mechanisms are reflected in the model by the application of the physics-based stress and strain invariants as input quantities. The application of the machine learning algorithm solved the problem of assigning an adequate parametric fatigue model to given material and loading conditions. The model was verified using the experimental data on the CuZn37 brass subjected to various cyclic loadings, including non-proportional multiaxial strain paths. The performance of the machine learning-based fatigue life prediction model is higher than the performance of the well-known parametric models.

Keywords: fatigue life prediction; CuZn37 brass; machine learning

Citation: Karolczuk, A.; Skibicki, D.; Pejkowski, Ł. Gaussian Process for Machine Learning-Based Fatigue Life Prediction Model under Multiaxial Stress–Strain Conditions. *Materials* **2022**, *15*, 7797. https://doi.org/10.3390/ma15217797

Academic Editor: Christian Motz

Received: 12 October 2022
Accepted: 3 November 2022
Published: 4 November 2022

1. Introduction

In the production and operation of machines, especially the means of transport, a constant effort to reduce costs and energy consumption is employed. This goal is achieved, inter alia, by weight reduction, which can be achieved by lowering safety factors, topology optimization, or by replacing traditional materials with alternative, less dense ones. Moreover, the design process departed from the infinite-life design strategy, which requires the stress to be lower than the fatigue limit. Increasingly, the machines are designed according to the safe-life design strategy for the durability estimated by the designer. When the repair is very expensive (e.g., jet engines), a damage-tolerant design strategy is applied. Here, the service of a machine with diagnosed damage was accepted [1]. Therefore, the design of machines and structures in accordance with the savings trends and the recommendations of the latest strategies requires accurate fatigue life estimation methods accompanied by an uncertainty estimation of predictions.

The drive to reduce costs also applies to the design process. For example, the use of computational models should not require expensive experimental studies to obtain additional and more reliable material data. The high expenditure especially regards the design against material fatigue because the fatigue damage process is difficult to model. First, it consists of many fatigue crack development periods, such as crack nucleation, small crack growth, and macroscopic crack growth, and these periods differ in physical damage mechanisms also influenced by size and notch effects [2]. Each period is affected by different driving forces and requires different approaches, such as the crack nucleation and crack growth approaches [3]. Second, the fatigue behavior of machines and structures generally exhibits spectacular stochastic behavior [4–6]. This results from the variability of fatigue loadings, parts geometry, material properties, and microstructures. Such a significant level of difficulty in fatigue assessment has led to the development of new fatigue models, despite the numerous models that have been developed in the past few decades [7–10].

These models simplified the complexity of material fatigue by using empirical or semi-empirical approaches to relate the primary loading quantities and material properties with the fatigue life. The empirical models adjust the parameters of the regression equation to the experimental data. Semi-empirical models combine fundamental physical principles and an empirical approach. Despite the use of some physical foundations to formulate models, the selected stress/strain quantities and material parameters are related by arbitrarily formulated parametric functions to obtain the best fit for the results of the experiment.

It is concluded that in the design process of engineering structures, the selection of the fatigue model is the primary problem. The choice should theoretically depend on the material properties (e.g., brittle or ductile), the type of loading (uniaxial or multiaxial, deterministic or random, proportional or non-proportional, stress ratio), and the range of deformation (elastic or elastic-plastic). In practice, owing to the multiplicity and complexity of models, the choice is most often limited by the level of the designer's knowledge. Conversely, owing to the cost and time-consuming nature of fatigue tests, the choice also results from the limited availability of data on the material and loading.

Based on the briefly presented problems of fatigue life prediction and existing solutions, a machine learning (ML) approach is proposed as a substitute for the semi-empirical fatigue models. The main advantage of this model is that it does not require the selection of a parametric (predefined) form of the fatigue model. The fatigue ML-based model should self-accommodate existing data and correctly reflect the fatigue behavior for testing data. Among several ML approaches, the neural network (NN) is widely used [11–13] in the fatigue field. According to Chen and Liu [12], NN modeling requires a large number of data, and there is no standard procedure for selecting an optimal NN architecture. One of the alternative ML approaches is the Gaussian process (GP) for regression with unique features that favors its application in fatigue life prediction of materials.

- The GP-based model requires a considerably smaller sample size of training data than other ML techniques [14–16]. The current analysis successfully applied a sample size of 30. It is a size comparable to the sample size for the determination of two reference stress or strain fatigue curves [17].
- The inherent feature of the GP is the estimation of the probability distribution for the model outputs [18,19]. The probabilistic output of the GP is important for its application in fatigue life prediction. Owing to this feature, a conservative design of mechanical systems can be performed.
- The GP-based model for a limited dimension of the input data vector (it is considered that the maximum five-dimensional vector of input data would be necessary for life prediction, Section 2.4) is computationally very effective [20].
- The GP can estimate the relevance of each component of the input vector for effective prediction [18,20]. This feature allows us to interpret the influence of the selected input quantities on fatigue life.

Existing studies on the use of GP to predict the durability of engineering machines and structures can be classified into approaches based on pure correlation of measured signals with progressive degradation of mechanical systems and hybrid approaches, including physics-based quantities.

In the first approach, the tested signals are not related to any model of the damage mechanism. Therefore, the results of this approach have mainly practical and predictive but not explanatory purposes. For example, Huchet et al. [21] used environmental parameters, such as wind speed and direction, in the fatigue assessment of wind turbine structures. In the research of Aye et al. [22] and Hong et al. [23], the prediction of the remaining life of bearings was directly based on the acquired vibration signals. Mohanty et al. [24,25] applied online signals from piezoelectric sensors attached to selected aircraft components in the GP for fatigue crack growth prediction. Hirvoas et al. [26] applied the GP to reduce uncertainties in a wind turbine numerical model considering different input data as support and blade structural properties.

Hybrid data-driven models involve input quantities related to the recognition of the failure mechanisms of a system. Słoński [27] used the GP to identify concrete properties, among others, using minimal and maximal uniaxial stress values. Hu et al. [28] applied GP to estimate the uncertainty of fatigue crack growth in turbine discs in a study on fatigue crack growth evaluation. Ling and Mahadevan [29] proposed replacing computationally expensive finite element analysis for fatigue crack growth with the GP model. Farid [30] predicted fatigue failure under stochastic loading using a stress signal at the critical section of the mechanical component. The GP model was combined with an artificial neural network to enhance the predictive performance and provide uncertainty quantification.

In the briefly reviewed papers, the GP models were adjusted and trained for a given mechanical system. Consequently, the trained models cannot be applied to other mechanical systems, and thus, they can be mostly classified as health monitoring systems [31].

Karolczuk and Słoński [32] proposed a novel approach in which the GP operates as a multiaxial fatigue life prediction model. The normal and shear stress amplitudes on the critical plane were selected as the physics-based input quantities for the GP model with the application of the squared exponential covariance function. The model was successfully verified on S355N steel and 2124 T851 aluminum alloy under the cyclic proportional combination of bending and torsion loadings at the high cyclic fatigue regime (stress-based condition). The proposed novel approach for fatigue life prediction requires further research and validation, especially under multiaxial non-proportional loading and stress–strain conditions.

This research aims to validate the GP applied to build a multiaxial fatigue model for the life prediction of CuZn37 brass under proportional and non-proportional loadings and stress–strain conditions.

The scope of this research involved an overview of selected classic (parametric) fatigue life prediction models (Section 2), basic concepts (Appendix A), and covariance functions (Appendix B) of the GP. Next, the fatigue life was estimated with GP using different covariance functions. As a result of the calculations, the physical quantities of key importance for the fatigue process were selected and compared with the quantities indicated in the analysis of classic models. The GP results were validated by comparing the results with the results obtained using the parametric models proposed by Fatemi-Socie, Brown-Miller, Glinka et al., and Yu et al. The calculations were conducted for eight loading cases, that is, axial, torsional, combined proportional axial-torsion loading, and non-proportional loadings. The level of non-proportionality was different in the case of the applied loadings owing to the phase shift (90° out-of-phase) and various frequencies of the applied strains (four different asynchronous loadings).

2. Brief Review of Fatigue Life Prediction Models

Commonly applied semi-empirical fatigue models consist of fatigue damage parameters, which are scalar functions of spatial stress/strain components and reference fatigue curves. The fatigue damage parameter is used to reduce the spatial stress or strain state to a scalar quantity of the dimensions of stress, strain, or energy. It is then compared with the reference regression curves to calculate the fatigue life [33,34]. Uniaxial regression curves show an explicit relationship between the applied stress, strain, or energy values and the number of cycles to failure. These curves are the result of fatigue tests performed on a limited set of specimens and require the adoption of statistical assumptions [17]. Uncertainties in material parameters, loading and geometry can also be included in life prediction by the application of probabilistic modeling with sampling techniques [35–37].

The fatigue models differ in physical quantities, which were adopted as decisive factors for the fatigue process—predictors for the GP. For example, stresses, strains, loading non-proportionality factors, or strain energy, including elastic and plastic parts. These quantities, which fluctuate with time, are mostly reduced before incorporating them into the damage model to statistical parameters such as amplitudes, mean values, or maxima.

In this section, several selected models representing different physical principles are briefly described. A broader overview of the multiaxial fatigue models can be found in the papers concerning the critical plane approach [7], energy approach [38], or non-proportionality of loading [39].

2.1. Empirical Models

An example of a purely empirical model can be the "ellipse quadrant" proposed by Gough and Pollard for ductile materials [40,41]. Because this is an equation for a particular loading case, that is, torsion and bending, a model is a function of the applied shear τ_a and normal σ_a stress amplitudes, as follows:

$$\left(\frac{\tau_a}{t_{-1}}\right)^2 + \left(\frac{\sigma_a}{b_{-1}}\right)^2 = 1, \tag{1}$$

where t_{-1} and b_{-1} are the fatigue limits for fully reversed torsion and bending, respectively. For brittle materials, Gough proposed the "ellipse arc" equation as follows:

$$\left(\frac{\tau_a}{t_{-1}}\right)^2 + \left(\frac{b_{-1}}{t_{-1}} - 1\right)\left(\frac{\sigma_a}{b_{-1}}\right)^2 - \left(2 - \frac{b_{-1}}{t_{-1}}\right)\frac{\sigma_a}{b_{-1}} = 1. \tag{2}$$

The above equations define the fatigue limit state of the applied stress amplitudes. It can be developed for a state at an arbitrary number of cycles to failure, as proposed in [42]. Empirical models have limited application (only combined torsion and bending loading) because they are not consistent with invariant principles.

2.2. Stress Invariants Models

The two examples presented here are attempts to adopt the Huber–Mises yield criterion for fatigue by considering the hydrostatic stress. Both are a linear combination of the amplitude of the second deviator invariant J_2, and the mean or maximum value of the hydrostatic stress σ_H, as in the case of the Sines [43] or Crossland [44] model, respectively:

$$\sqrt{J_{2,a}} + k_s \cdot \sigma_{H,m} = f\left(N_f\right), \sqrt{J_{2,a}} + k_c \cdot \sigma_{H,max} = f\left(N_f\right), \tag{3}$$

where k_s, k_c are material parameters, $f\left(N_f\right)$ is the reference fatigue curve.

The fatigue models based on stress invariants are criticized mainly because of problems in their implementation of random loading and non-proportional loading [7,45]. To overcome these problems, special procedures must be implemented.

2.3. Critical Plane Models

Critical plane models are based on the observation of fatigue crack formation. Based on the observation, the fatigue cracks in metallic materials nucleate and develop in certain preferred planes within the material [46,47]. Thus, the critical plane approach assumes that the stress/strain components on a specific plane are primary for fatigue crack initiation and failure.

The stress-based Findley criterion is one of the first critical plane multiaxial fatigue models [48]. This is a linear combination of the maximum normal stress $\sigma_{n,max}$ and shear stress amplitude $\tau_{ns,a}$ on the plane for which the equation reaches its maximum, as follows:

$$max(\tau_{ns,a} + k \cdot \sigma_{n,max}) = f\left(N_f\right), \tag{4}$$

where k is a material parameter.

Brown and Miller [49] proposed a strain-based fatigue damage parameter composed of shear $\gamma_{ns,a}$ and normal $\varepsilon_{n,a}$ strain amplitudes on the plane experiencing the maximum shear strain amplitude. Kandil et al. [50] proposed a fatigue life prediction model based on this concept of the fatigue damage parameter, in the form

$$\gamma_{ns,a} + k_{BM} \cdot \varepsilon_{n,a} = f\left(N_f\right), \tag{5}$$

where k_{BM} is a material parameter.

Fatemi and Socie [51] proposed to relate the shear strain amplitude $\gamma_{ns,a}$ and maximum normal stress $\sigma_{n,max}$ normalized by the yield strength σ_{yield}, in the following form:

$$\gamma_{ns,a}\left(1 + k \cdot \frac{\sigma_{n,max}}{\sigma_{yield}}\right) = f\left(N_f\right). \tag{6}$$

Carpinteri et al. [52] proposed a nonlinear function of amplitude and mean value of normal stress and shear stress amplitude on the critical plane related to the average principal stress directions in the following form:

$$\sqrt{\left(\sigma_{n,a} + a_{CS}\sigma_{n,m}\right)^2 + b_{CS}\tau_{ns,a}^2} = f\left(N_f\right), \tag{7}$$

where a_{CS}, b_{CS} are material parameters.

Papuga–Růžička [53] also proposed a nonlinear function of amplitude and mean value of normal stress and shear stress amplitude on the critical plane of its maximum, as follows

$$\max_{n}\left\{\sqrt{a_{PR}\tau_{ns,a}^2 + b_{PR}\left(\sigma_{n,a} + c_{PR}\sigma_{n,m}\right)}\right\} = f\left(N_f\right), \tag{8}$$

where a_{PR}, b_{PR}, c_{PR} are material parameters. The proposed formula was developed [45,54] to take into account the mean shear stress effect.

Glinka, Shen, and Plumtree [55] proposed the strain energy parameter accounting for both strains and stresses in the plane of maximum shear strain, as follows:

$$\gamma_{ns,a}\tau_{ns,a} + \varepsilon_{n,a}\sigma_{n,a} = f\left(N_f\right). \tag{9}$$

This model was modified by Pan-Chun-Chen [56] by introducing a weighting factor k_G to normal components, as follows:

$$\gamma_{ns,a}\tau_{ns,a} + k_G\varepsilon_{n,a}\sigma_{n,a} = f\left(N_f\right). \tag{10}$$

Ince and Glinka [57] introduced a generalized strain energy fatigue damage parameter as a function of the elastic and plastic strain energy density contributed by the normal and shear stresses and strains on the critical plane of its maximum in the following form:

$$\max_{\mathbf{n}}\left(\tau_{ns,a}\gamma_{ns,a}^{e} + \tau_{ns,a}\gamma_{ns,a}^{p} + \sigma_{n,max}\varepsilon_{n,a}^{e} + \sigma_{n,max}\varepsilon_{n,a}^{p}\right) = f\left(N_f\right), \tag{11}$$

where $\mathbf{n}$ is a unit vector that determines the orientation of the plane.

Yu et al. [58] modified the Ince-Glinka parameter and proposed the following model:

$$\frac{\gamma_{ns,a}\tau_{ns,a}}{\tau_f'} + \frac{2\varepsilon_{n,a}\sigma_{n,max}}{\sigma_{yield} + \sigma_f'} = f\left(N_f\right), \tag{12}$$

computed on the plane of maximum shear strain, where τ_f' and σ_f' are material parameters deduced from the uniaxial reference curves.

2.4. Summary and Model Selection

It is assumed that the GP-based fatigue model (Appendix A) can substitute any parametric functions proposed for fatigue life prediction. The substitution should be effective because the GP can map any fatigue behavior deduced from training data, and thus, the selection problem of adequate parametric function, for example, Equations (1)–(12), are omitted. The effectiveness of the GP-based model is the highest if the input data vector for the GP includes fatigue damage-related quantities. Based on the effort of many researchers and their attempt to fit the experimental data to parametric models, the primary quantities (predictors for the GP) for fatigue damage of metallic materials can be selected from the shortlist presented in Table 1. The five predictors were selected from the critical plane models as being consistent with the observed fatigue damage mechanism of metallic materials. The role of these quantities and their interaction are discussed in [59,60]. The quantities from stress-invariant models could also be considered; however, owing to the ambiguous definition of amplitudes for stress invariants [61], they were neglected in the present research. Stress-based quantities are commonly applied at high cyclic fatigue (HCF), whereas strain-based quantities are common at low cyclic fatigue (LCF). Under non-proportional loading, the principal stresses rotate, which activates the higher number of slip systems and induces the complex interaction between dislocation movements. However, applying both quantities, i.e., strain and stress, the above mechanisms are reflected in predictors and the machine learning model is able to recognize this pattern to more effectively predict the fatigue life. The application of stress-and strain-based quantities reflects any possible additional material hardening effects occurring under non-proportional loading for some metallic materials. In the present research, we consider the plane or maximum shear strain amplitude appropriate for a wide class of metallic materials with dominant shear or mixed shear/tensile damage mechanisms.

Table 1. Primary predictors for fatigue life of metallic materials.

No	Predictor	Description
1	$\gamma_{ns,a}$	Shear strain amplitude—primary parameter at LCF for materials with dominant micro shear cracking
2	$\tau_{ns,a}$	Shear stress amplitude—primary parameter at HCF for materials with dominant micro shear cracking
3	$\varepsilon_{n,a}$	Normal strain amplitude—primary parameter at LCF for materials with dominant tensile cracking or materials with mixed shear/tensile cracking
4	$\sigma_{n,a}$	Normal stress amplitude—primary parameter at HCF for materials with dominant tensile cracking or materials with mixed shear/tensile cracking
5	$\sigma_{n,m}$	Mean value of normal stress—primary parameter at HCF for materials with dominant tensile cracking or materials with mixed shear/tensile cracking. It reflects the beneficial effect of compressive mean stress for fatigue

The fatigue life prediction performance of the proposed GP-based model was compared with the performance of four parametric fatigue models, including the proposal of Brown–Miller (5), Fatemi–Socie (6), Glinka et al. (8), and Yu et al. (10). The selected models required a more detailed description of the implemented reference fatigue curves $f\left(N_f\right)$ and weighting material parameters. The basic regression curve implemented in defining the reference curve for each model is based on the Manson–Coffin curve [62] under fully reversed cyclic torsion loading, as follows:

$$\gamma_f\left(N_f\right) = \frac{\tau_f'}{G}\left(2N_f\right)^{b_0} + \gamma_f'\left(2N_f\right)^{c_0}, \tag{13}$$

where τ_f' is the shear fatigue strength coefficient, G is the shear modulus, b_0 is the shear fatigue strength exponent, γ_f' is the shear fatigue ductility coefficient, and c_0 is the shear fatigue ductility exponent. The Brown–Miller, Fatemi–Socie, and Glinka et al. models include the weighting material parameters that can have a life-dependent form [33,63] to be fully consistent with two uniaxial strain-life curves obtained under tension-compression and torsion loadings. The torsion strain-life curve is given by Equation (13), and the tension-compression strain-life curve is

$$\varepsilon_f\left(N_f\right) = \frac{\sigma_f'}{E}\left(2N_f\right)^{b} + \varepsilon_f'\left(2N_f\right)^{c}, \tag{14}$$

where σ'_f is the axial fatigue strength coefficient, E is Young's modulus, b is the axial fatigue strength exponent, ε'_f is the axial fatigue ductility coefficient, and c is the axial fatigue ductility exponent. The first component of Equations (13) and (14) are the elastic strain and plastic strain, respectively. Decomposition into elastic and plastic parts is consistent with the physical mechanism, and it is necessary to derive life-dependent material weighting factors appropriately. This decomposition leads to the following notations:

$$\varepsilon^e_f\left(N_f\right) = \frac{\sigma'_f}{E}\left(2N_f\right)^b,\ \varepsilon^p_f\left(N_f\right) = \varepsilon'_f\left(2N_f\right)^c,\ \sigma_f\left(N_f\right) = \sigma'_f\left(2N_f\right)^b,\ \tau_f\left(N_f\right) = \tau'_f\left(2N_f\right)^{b_0}, \tag{15}$$

where the upper indexes p and e indicate the plastic and elastic strain parts, respectively. Implementing these notions, the detailed formulas for the parametric life prediction models were obtained [63], as follows:

- Brown–Miller model:

$$f\left(N_f\right) = \gamma_f\left(N_f\right) \tag{16a}$$

$$k_{BM}\left(N_f\right) = \frac{2\left[\gamma_f\left(N_f\right) - (1+\nu^e)\varepsilon^e_f\left(N_f\right) - (1+\nu^p)\varepsilon^p_f\left(N_f\right)\right]}{(1-\nu^e)\varepsilon^e_f\left(N_f\right) + (1-\nu^p)\varepsilon^p_f\left(N_f\right)}, \tag{16b}$$

where ν^e and ν^p are the elastic and plastic Poisson ratios, respectively.

- Fatemi–Socie model:

$$f\left(N_f\right) = \gamma_f\left(N_f\right) \tag{17a}$$

$$k_{FS}\left(N_f\right) = \left[\frac{\gamma_f\left(N_f\right)}{(1+\nu^e)\varepsilon^e_f\left(N_f\right) + (1+\nu^p)\varepsilon^p_f\left(N_f\right)} - 1\right]\frac{2\sigma_{yield}}{\sigma_f\left(N_f\right)} \tag{17b}$$

- Glinka et al. model:

$$f\left(N_f\right) = \gamma_f\left(N_f\right)\tau_f\left(N_f\right) \tag{18a}$$

$$k_G\left(N_f\right) = \frac{4\gamma_f\frac{\tau_f\left(N_f\right)}{\sigma_f\left(N_f\right)} - 2\left((1+\nu^e)\varepsilon^e_f\left(N_f\right) + (1+\nu^p)\varepsilon^p_f\left(N_f\right)\right)}{(1-\nu^e)\varepsilon^e_f\left(N_f\right) + (1-\nu^p)\varepsilon^p_f\left(N_f\right)} \tag{18b}$$

- Yu et al. model:

$$f\left(N_f\right) = \gamma'_f\left(2N_f\right)^{c_0} + \frac{\tau'_f}{G}\left(2N_f\right)^{2b_0}. \tag{19}$$

Substituting Equations (13)–(19) in (6) and (8)–(10) and implementing an iterative gradient-based procedure results in the calculation of the fatigue life N_{cal} for each parametric model.

3. Experiment

Tubular unnotched thin-walled specimens made of CuZn37 brass were subjected to fully reversed constant-amplitude fatigue loading under various loading paths. Because the experimental test details can be found in [64], only the most important information is reported. The strain-controlled fatigue tests were conducted according to the ASTM E2207-02 standard. The failure definition was a 10% drop in the axial force or torque. The monotonic and cyclic mechanical properties of the CuZn37 brass are listed in Table 2. The regression fatigue curves (13) and (14) are shown in Figure 1. The Monte Carlo sampling technique was applied to generate the distribution of the fatigue curves ((*i*)-indexed curves in Figure 1) using a multivariate normal distribution of regression coefficients [36,65]. A sample size of 1000 was applied, estimated from the stability analysis of error indexes (Section 4.1).

The blue vertical dashed lines in Figure 1 indicate the overlapping experimental fatigue life regimes of the axial and shear fatigue curves. The experimental setup included eight different loading paths with a fixed strain ratio of amplitudes $\gamma_{xy,a}/\varepsilon_{xx,a}$. Each loading path was tested using 10 specimens at different strain loading amplitudes. The shapes of the loading paths with applied strain ratios are presented in Figure 2a. The material responses in the form of stress paths are displayed in Figure 2b. The stress amplitudes implemented in the fatigue life prediction were determined for the half-life. The received experimental fatigue lives $2N_{exp}$ were within the range of [383, 191000] reversals. The experimental data file with the registered signals (strain and stress components) was uploaded to the Mendeley data repository (https://data.mendeley.com/datasets/7fbkf6y7gv/1, accessed on 25 October 2021).

Table 2. Mechanical properties of CuZn37 brass.

E (GPa)	ν^e	ν^p	G (MPa)	σ_u (MPa)	σ_{yield} (MPa)	K' (MPa)	n'
105	0.33	0.50	39.5	366	138	819	0.2142
γ_f'	c_0	τ_f' (MPa)	b_0	ε_f'	c	σ_f' (MPa)	b
0.5065	−0.4370	204	−0.0475	0.3853	−0.5269	393	−0.0526

K', n' are cyclic strength coefficient and strain hardening exponent, respectively.

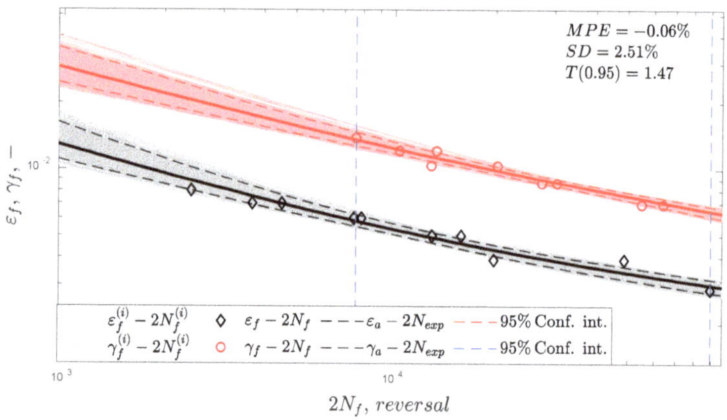

Figure 1. Shear and axial strain-life fatigue curves for CuZn37 with error indexes $MPE, SD, T(0.95)$ (described in Section 4.1).

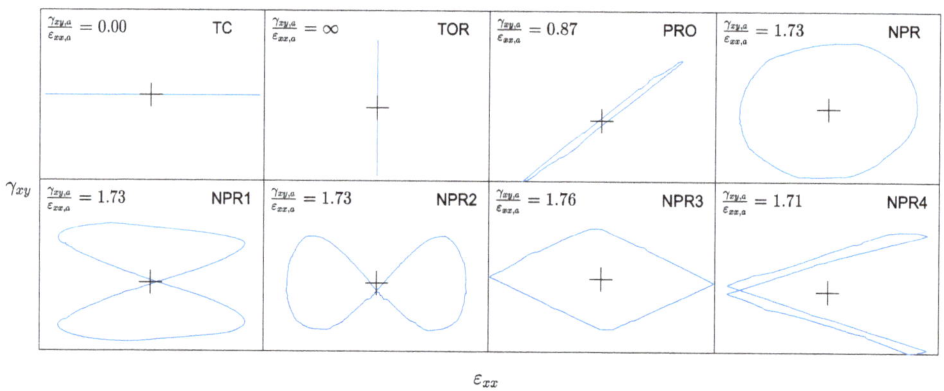

(**a**)

Figure 2. *Cont.*

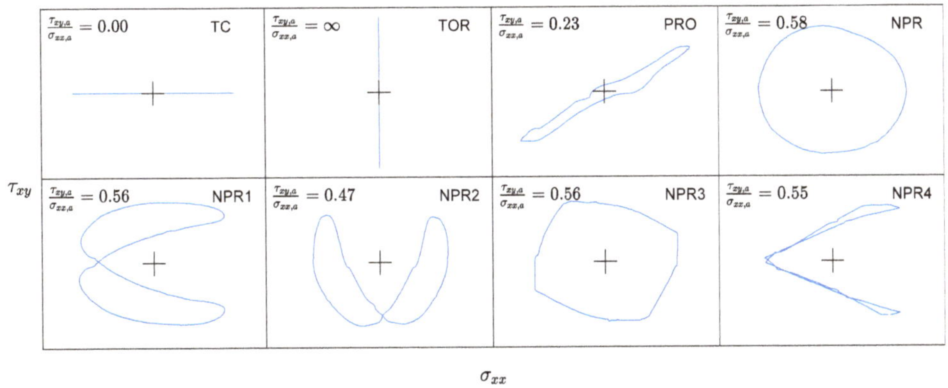

(b)

Figure 2. Registered experimental strain paths in (a) axial-shear strain space, $\varepsilon_{xx} - \gamma_{xy}$ and (b) axial-shear stress space, $\sigma_{xx} - \tau_{xy}$.

4. Results and Discussion

The normal and shear stress/strain components on the critical plane implemented in the parametric (Section 2) and GP-based fatigue life prediction (Appendix A) models were calculated based on the registered strain and stress tensor components. Searching for the plane orientation with the maximum shear strain (the critical one), numerical simulations were conducted with the application of Euler angles, φ, θ—describing the plane orientation [66]. In the simulation, the step between subsequent values of the Euler angles was equal to 1°. For the non-proportional loading paths, instability in the calculated fatigue lives was detected as a result of the existence of different planes with the same maximum value of shear strain but with different normal stress and strain components. To overcome this problem, the critical plane was searched within the boundary $\langle 1-\delta, 1\rangle \max_{\varphi,\theta} \gamma_{ns}$, where δ = 0.0001 was applied. Within this boundary, the maximum values of normal stress, normal strain, and shear stress were determined and applied to the fatigue life calculations (the values could be found in the file uploaded to https://data.mendeley.com/datasets/7fbkf6y7gv/1, accessed on 25 October 2021).

4.1. Error Indexes

The calculated fatigue lives were compared with the experimental fatigue lives to estimate the performance of each applied fatigue model. A few error indexes were implemented in which the first group belongs to purely statistical parameters of fitting, and the second group belongs to parameters to estimate the applicability of models to life prediction. The first group is based on a percentage error, defined as

$$PE = \frac{logN_{exp} - logN_{cal}}{logN_{exp}} \times 100, \quad (20)$$

where N_{exp} and N_{cal} are the experimental and calculated number of cycles to failure, respectively. The statistics of PE provide information on the efficiency of the model in fatigue life prediction. The mean values, MPE and standard deviation, SD of PE were evaluated. The positive values of PE concern cases where the calculated life was shorter than the experimental one (conservative estimate), whereas non-conservative estimates were indicated by negative PE values. An MPE value equal to zero indicates a perfect prediction-unreal case owing to fatigue life scatter.

The mean and standard deviation of PE estimate the fitting properties of the applied model. However, to build (based on these parameters) the ranking of models with respect to their acceptable uncertainty in life prediction can be questionable. First, the two parameters are combined into a decisive one. Second, the values of these parameters are not intuitive if the predicted fatigue lives are acceptable. The experimental fatigue life can vary significantly even under a single loading condition [67,68]. The fatigue scatter factor is defined as:

$$T = \begin{cases} \frac{N_{exp}}{N_{cal}} & for \quad N_{exp} \geq N_{cal} \\ \frac{N_{cal}}{N_{exp}} & for \quad N_{exp} < N_{cal} \end{cases} \tag{21}$$

Fatigue scatter factor less than three is commonly accepted [69,70], and a value less than or equal to two exhibits a very good estimation [71–73]. Thus, the statistical distribution of the T factor was used to define [74] a more intuitive error index as a 0.95-quantile of the fatigue life scatter factor. The 0.95-quantile ($T(0.95)$) was calculated based on the shape-preserving piecewise cubic interpolation of the empirical cumulative distribution of T. The value of $T(0.95)$ determines the minimum fatigue-life scatter band required to include 95% of all experimental fatigue data.

4.2. The Parametric Fatigue Life Prediction Models

A comparison of the experimental $2N_{exp}$ and calculated $2N_{cal}$ fatigue lives for the four analyzed parametric models is presented in Figure 3. Each panel in Figure 3 includes the solid line of perfect life consistency enclosed by dashed lines of the scatter band $T = 2.0$. The error bars indicate 95% prediction intervals computed using the Monte Carlo sampling technique on strain-life fatigue curves. Additionally, the results for different loading paths are marked, and the corresponding values of the error indexes are included in the legend of the figure panels. The data used for the calibration of the model, i.e., uniaxial tension–compression and pure torsion were labeled as 'Train data' and the remaining data as 'Test data'. For the labeled data, the MPE, SD, and $T(0.95)$ error indexes were estimated and presented in each panel.

The error indexes are identical for the analyzed models of Fatemi–Socie, Brown–Miller, and Glinka et al., with the application of the life-dependent material parameters that equalize the performance of the parametric models under data used for calibration. The obtained error indexes $MPE = -0.1\%$, $SD = 2.5\%$, and $T(0.95) = 1.5$ for the training data and the models characterize the experimental life scatter for the CuZn37 brass under the fatigue test conditions. These values were treated in further analyses as the reference consistency of fatigue lives. Yu et al. applied constant weighting factors of shear and normal strain energy parameters, and the results for the analyzed CuZn37 brass were ineffective even under uniaxial and pure torsion loadings with $MPE = -23.7\%$ and $MPE = -9.8\%$, respectively (Figure 3d).

The fatigue lives for the multiaxial proportional loading path were calculated using models with life-dependent material parameters within the acceptable value of error indexes (80% of the data is included within the scatter T = 2.0, and 100% of the data is included within the scatter T = 3.0). For the non-proportional loading paths, the best life prediction performance was exhibited using the Fatemi–Socie model (Figure 3a). However, the fatigue life under loading path NPR4 was unsuccessfully estimated with $MPE = 7.7\%$. The performances of the models by Brown–Miller and Glinka et al. were the worst, with MPE values exceeding 14% and 27% for the NPR4 path, respectively. The Brown–Miller model with shear strain γ_{ns} linearly combined with the normal strain ε_n resulted in a very conservative life estimation under non-proportional loading paths (Figure 3b). This means that the non-proportional loading paths led to a higher value of the normal strain on the critical plane of the maximum shear than expected using the Brown–Miller model. The same effect, but magnified by the shear stress on the critical plane, was observed in the model by Glinka et al.

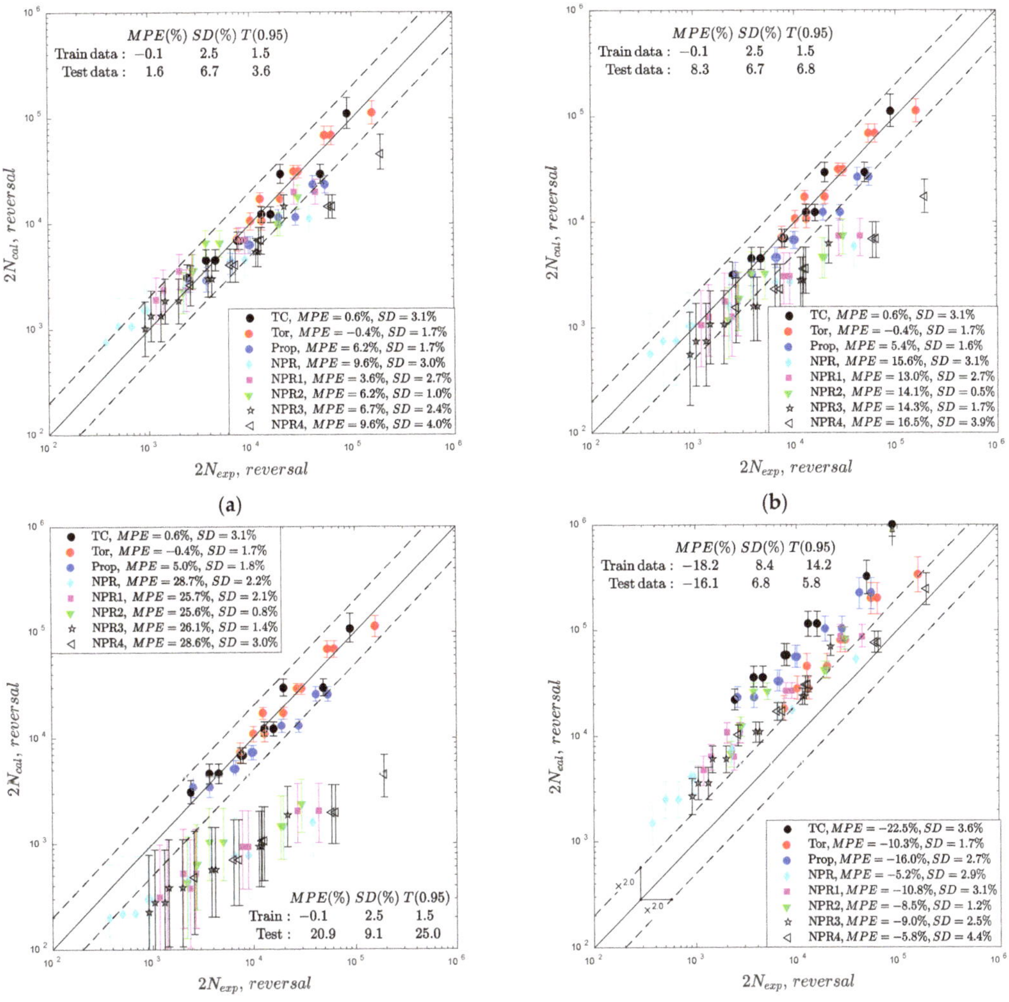

Figure 3. Comparison of experimental and calculated fatigue lives for the models of (**a**) Fatemi–Socie, (**b**) Brown–Miller, (**c**) Glinka et al., and (**d**) Yu et al.

The tendency of higher life underestimation ($N_{cal} < N_{exp}$) for longer experimental fatigue lives was observed for all models with the life-dependent material parameters under non-proportional loading paths. It is concluded that the CuZn37 brass experienced additional hardening under non-proportional loading paths that resulted in higher stress values than expected by the models, which were calibrated by uniaxial and pure torsion loading paths. Yu et al. overestimated ($N_{cal} > N_{exp}$) the predicted fatigue lives for all loading paths.

4.3. The Gaussian Process-Based Fatigue Model

4.3.1. Physics-Based Input Parameters (Predictors)

The GP model is assumed to substitute the semi-empirical fatigue criteria; thus, the scalar output of the logarithm of the fatigue life must be invariant under the rotation of coordinate systems. To meet this requirement, the covariance function for the GP must not directly implement the input parameters as stress and/or strain tensor components. The rotation of the coordinate system transforms the stress/strain tensor components, and thus, the model trained on the dataset valid for the original coordinate system would not apply to the rotated coordinate system.

A state-of-the-art review of fatigue life prediction models (Section 2) indicates a few commonly applied crucial physics-based quantities. These are derived from the concept of the critical plane built on the experimental observation that fatigue cracks in metals are initiated in the plane of maximum shear stress or strain [49,75], τ_{ns}, γ_{ns}. Furthermore, these microcracks could not be developed if the maximum normal stress or strain σ_n, ε_n on this plane were below the critical value. These four parameters are considered primary in most multiaxial fatigue life prediction models (Table 1). Owing to the zero mean stresses in the experimental data, the $\sigma_{n,m}$ predictor is omitted for components of the input data vector x, as follows:

$$x = [\gamma_{ns}, \varepsilon_n, \tau_{ns}, \sigma_n]. \tag{22}$$

The maximum shear strain plane was determined over the entire loading history and fixed in the fatigue life calculation process. Thus, the selected predictors (22) are invariant under the rotation of the coordinate system. Regarding the identification of multiple planes with equal maximum values of shear strain, the plane with the highest normal stress was selected as the critical plane. The relevance of each predictor was analyzed during the training process, as presented in the next section. The diagram of data flow for the fatigue life prediction based on the proposed GP fatigue model is presented in Appendix A, Figure A1.

4.3.2. Training Process

The polycrystalline structure with preferred slip systems makes metallic materials sensitive to the type of fatigue loading paths. For example, the non-proportional loading characterized by the rotation of principal stresses could activate a larger number of slip systems and their intensive interactions compared to the proportional loading path [76]. These phenomena could influence the fatigue life of materials; however, this effect depends on many factors [77,78]. If the fatigue GP-based model is expected to be implemented under non-proportional loading for materials sensitive to its effect, the training data should include such a case. The analyzed experimental data included various non-proportional loading paths (Section 3, Figure 2), and the most common one (easy to replicate) with a 90° phase shift (Path NPR in Figure 1) was selected for inclusion in the training process. To analyze the effect of the non-proportional loading path on the fatigue life, two training datasets were implemented. The first training dataset includes only loading paths commonly applied to identify the parametric fatigue life prediction models (Section 2), that is, uniaxial push-pull (10 specimens) and pure torsion (10 specimens) loadings. The second training dataset additionally included the NPR path (10 specimens).

Initially, the training process included all four predictors (22), five specified covariance functions (Appendix B), and the first training dataset. The length scales l_i for each covariance function and each predictor were used to calculate the relevance factor (RF), defined as:

$$RF = \left(\frac{l}{std(x)}\right)^{-1}, \tag{23}$$

where l is the length scale, and $std(x)$ is the standard deviation of the analyzed predictor observed in the training data. The larger the length-scale parameter, the lower the covariance, and thus the lower the influence of its predictor. However, it also depends on the analyzed physic-based quantity; for example, the strains in the fatigue regime are several orders of magnitude below the stresses. This can be considered by normalizing the length scale using the standard deviation of the input of the analyzed predictor, $std(x)$. The reverse of $l/std(x)$ is defined as the RF. An RF value approaching zero indicates that the analyzed predictor can be ignored.

The RFs obtained for the first training dataset are shown in Figure 4. The vector of the standard deviations for the first training dataset is $\text{std}(x) = [0.0026\ (-),\ 0.0009\ (-),\ 6.6\ (\text{MPa}),\ 62\ (\text{MPa})]$. Two conclusions can be drawn. First, the exponential covariance function (EX) exhibits the lowest values of the relevance factors compared to other kernels. This means that the predicted values for the test data applying the EX kernel will be characterized by a larger uncertainty than for the remaining kernels. Second, the predictor ε_n—of the normal strain on the critical plane can be neglected for M5/2, RQ, and SE kernels (near zero relevance factor). The mean percentage error (MPE), standard deviation of (PE), and $T(0.95)$ factor for all the covariance functions are in the ranges $MPE = [-0.07, -0.06]\ \%$, $SD = [2.0,\ 2.5]\ \%$, $T(0.95) = [1.53,\ 1.54]$. Based on the obtained results, further analysis was reduced to four covariance functions (M3/2, M5/2, RQ, and SE) and three predictors, $x = [\gamma_{ns}, \tau_{ns}, \sigma_n]$. The error-fitting indicators; MPE, SD, and $T(0.95)$ were in the same range as the initial selection. The estimated values of the hyperparameters for the analyzed kernels for the first training dataset are presented in Table 3.

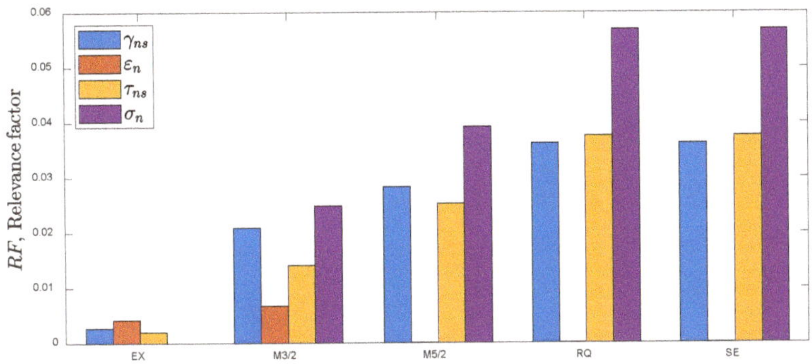

Figure 4. RFs obtained by application of different kernels for the first training dataset (uniaxial and torsion).

Table 3. Hyperparameters of the covariance functions for the first training dataset.

Kernel	Length-Scales			Scale-Mixture Parameter	Standard Deviation of the Noise	Standard Deviation of the (Noise-Free) Signal
	l_γ, (-)	l_τ, (MPa)	l_σ, (MPa)	α, (-)	σ_y, (-)	σ_k, (-)
M3/2	0.1239	480.5	2430	-	0.1099	6.74
M5/2	0.0921	262.8	1581	-	0.1106	6.12
RQ	0.0720	176.5	1091	1.527×10^5	0.1109	5.75
SE	0.0720	176.5	1091	-	0.1109	5.75

The relevance factors for the second training dataset, including the non-proportional loading path NPR, are presented in Figure 5. Based on the previous analysis, estimation was conducted on four covariance functions and three predictors. Here, the vector of standard deviations of the predictors was found as $\text{std}(x) = [0.0027(-),\ 22.4(\text{MPa})\ 124(\text{MPa})]$. For the second training dataset, all four kernels indicate the irrelevance of the shear stress τ_{ns} on the critical plane. Thus, in the final selection, the shear stress as a predictor was abandoned. The error fitting indicators for both selections are in the ranges of $MPE = [-0.07, -0.06]\ \%$, $SD = [2.15,\ 2.50]\ \%$, $T(0.95) = [1.44,\ 1.67]\ \%$.

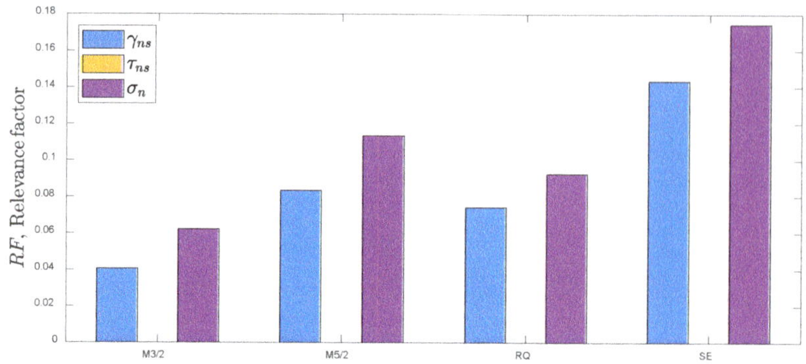

Figure 5. RFs obtained by application of different kernels for the second training dataset (uniaxial, torsion, and NPR path).

Inclusion of the non-proportional loading path NPR into the training data increased the RFs for all kernels by approximately four times. The estimated values of the hyperparameters for the analyzed kernels for the second training dataset are presented in Table 4.

Table 4. Hyperparameters of the covariance functions for the second training dataset.

Kernel	Length Scales		Scale-Mixture Parameter	Standard Deviation of the Noise	Standard Deviation of the (Noise Free) Signal
	l_γ, (-)	l_σ, (MPa)	α, (-)	σ_y, (-)	σ_k, (-)
M3/2	0.0657	1996	-	0.1005	6.16
M5/2	0.0321	1094	-	0.1059	5.40
RQ	0.0361	1344	0.1301	0.1079	6.00
SE	0.0187	712.0	-	0.1096	5.83

4.3.3. Test Process

In contrast to the parametric models in which estimation of output uncertainty requires additional methodologies, for example, Monte Carlo sampling [36], the inherent property of the GP model is the variance estimation of the outputs. This property was utilized in the fatigue life prediction shown in Figure 6 by additional vertical bars providing 95% prediction intervals. Figure 6 presents a comparison of the experimental and calculated fatigue lives obtained by the implementation of M3/2, M5/2, RQ, and SE kernels trained on the first dataset (only tension-compression and pure torsion paths).

The results demonstrated that within the test data (6 loading paths and 60 specimens), only fatigue lives predicted for the multiaxial proportional loading can be accepted with $MPE = [3.7, 4.2]\%$ and all data included within the fatigue scatter band $T = 3.0$. The fatigue lives under the non-proportional loading paths with no exception are predicted with high underestimation with $MPE = [22.8, 37.6]\%$ and $T(0.95) = [13.2, 88.4]$. Additionally, the 95% prediction intervals are approximately two and five times larger than those estimated for the training data with the implementation of the M3/2 and M5/2 kernels. A summary of the error indexes received for the analyzed kernels is presented in Figure 7. The lowest error index was obtained for the M3/2 kernel, but with high (95%) prediction intervals and unacceptable values of $T(0.95) > 3$.

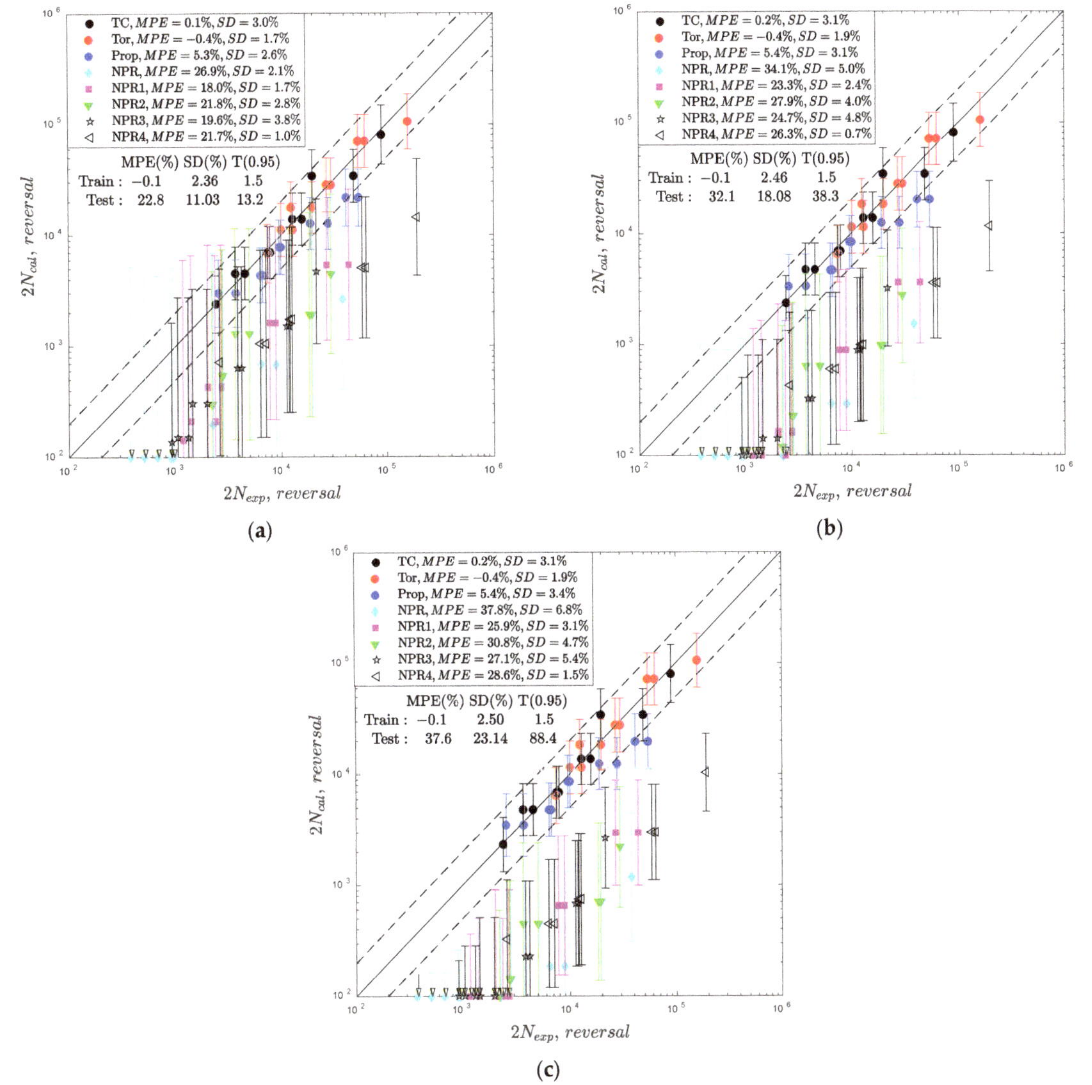

Figure 6. Comparison of experimental and calculated fatigue lives for the fatigue GP-based model for the first training dataset and the following covariance functions: (**a**) Matern 3/2, (**b**) Matern 5/2, (**c**) RQ/SE.

It is concluded that the fatigue GP-based model trained on the dataset without the non-proportional loading paths cannot predict the material behavior of CuZn37 brass. The non-proportional loading induced an additional fatigue phenomenon in the CuZn37 brass compared to the phenomenon induced under proportional loading.

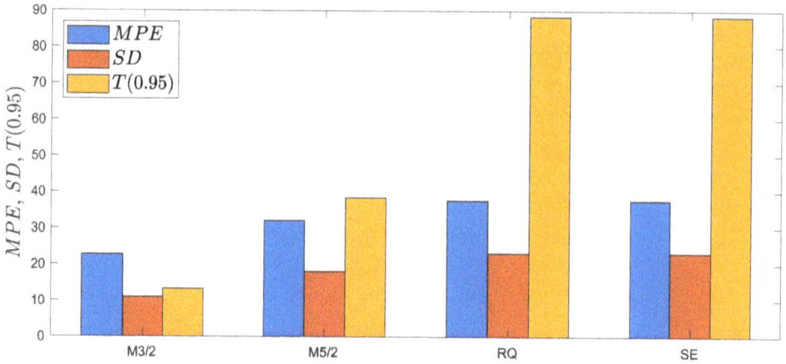

Figure 7. Comparison of error indexes obtained by application of different kernels to the fatigue GP-based model trained on the first training dataset.

The results obtained for the fatigue GP-based model trained on the second dataset, including one non-proportional loading path NPR, are displayed in Figure 8. By adding the results for one non-proportional loading path with 10 specimens to the training data, the prediction performance of the GP model was improved. An improvement is observed for all the applied covariance functions. The error indexes for the test dataset and analyzed kernels are practically equal (Figure 9) with $MPE = [-0.2, -0.1]\%$ and $T(0.95) = [2.4, 2.5]$.

The training dataset with 30 specimens successfully trained the GP model for fatigue life prediction of CuZn37 brass subjected to proportional and non-proportional loading paths with test data of 50 specimens. One standard non-proportional loading path with a 90° phase shift provided sufficient information on the fatigue behavior of the GP model. Consequently, the fatigue life was successfully predicted for the four tested non-proportional loading paths (NPR1, NPR2, NPR3, NPR4) of different and complex shapes (Figure 2) and for one proportional loading path (Prop). Furthermore, the common complex measures of the non-proportional loading effect on fatigue life, such as non-proportional factors [79–82], integral approaches [66,72,82] and enclosing surface methods [83,84], are not needed.

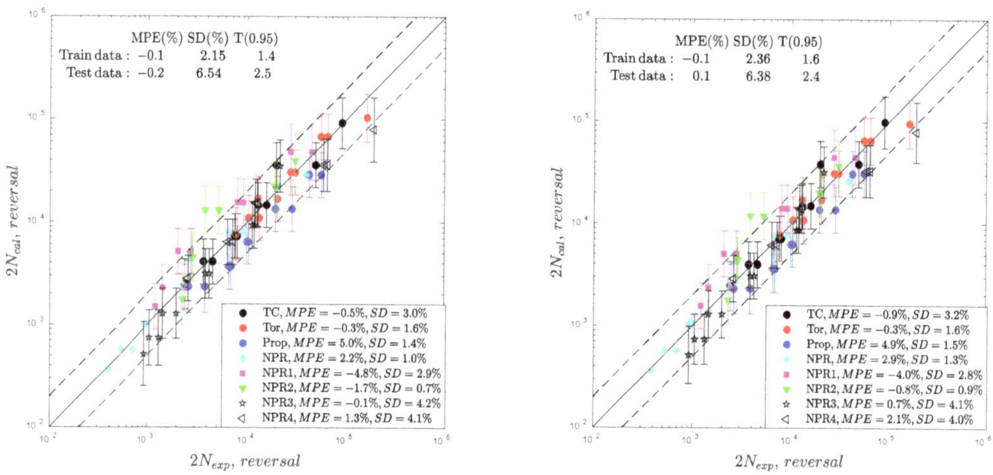

(a) (b)

Figure 8. *Cont.*

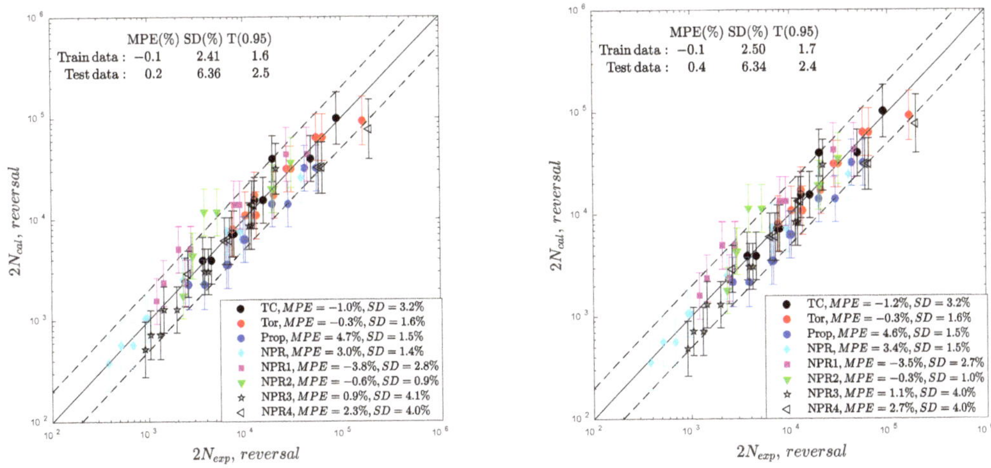

(c) (d)

Figure 8. Comparison of experimental and calculated fatigue lives for the fatigue GP-based model for the second training dataset and the following covariance functions: (**a**) Matern 3/2, (**b**) Matern 5/2, (**c**) RQ, and (**d**) SE.

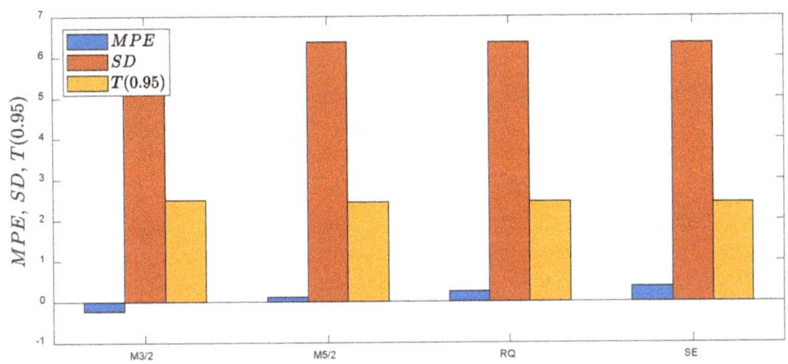

Figure 9. Comparison of error indexes obtained by the application of different kernels to the fatigue GP-based model trained using the second dataset.

5. Summary and Conclusions

The Gaussian process (GP) was applied to build the fatigue model for the life prediction of CuZn37 brass subjected to various multiaxial loading paths. The physics-based input quantities in the form of stress and strain components on the critical plane of maximum shear served as predictors in the GP-based model. These quantities are invariant under the rotation of the coordinate system, and thus, the trained fatigue GP-based model is consistent with the invariance principles. Five stationary covariance functions for the GP with length scales assigned to each predictor (anisotropic kernels) were implemented and analyzed. Two Matern class kernels (M3/2, M5/2), SE, and RQ were preliminarily accepted for final validation based on the relevance analysis of predictors. The predictive performance of the GP-based model was verified using two training datasets. The first training dataset included only uniaxial and pure torsion loading paths, and the second dataset included one non-proportional loading path. The detailed conclusions are as follows.

- The relevance analysis of the applied input quantities for the fatigue GP-based model revealed that the maximum shear strain and normal stress on the plane of maximum shear are the most decisive factors for the life prediction of CuZn37 brass.

- The GP model trained on uniaxial and pure torsion loading paths was ineffective for the prediction of the fatigue life of CuZn37 brass under non-proportional loading paths.
- Two Matern-class kernels (M3/2, M5/2), the SE kernel, and the RQ kernel were successfully applied to the GP-based model with better prediction performance than the parametric commonly applied multiaxial criteria of Fatemi–Socie, Brown–Miller, Glinka et al., and Yu et al.
- The computational time was decreased approximately 7.8 times by applying the GP-based model compared to the parametric fatigue models.
- The effect of mean loading can be simply implemented in the proposed fatigue GP-based model by adding the mean components of stress/strain to the input quantities (predictors).

The GP-based model can effectively substitute parametric fatigue life prediction models if physics-based predictors consistent with invariant principles are applied. Further validation for different types of materials under the mean stress effect and a wider fatigue life regime is needed.

Author Contributions: A.K.: conceptualization, methodology, software, data curation, investigation, writing—original draft, writing—review and editing, visualization. D.S.: investigation, resources, writing—review and editing. Ł.P.: investigation, resources, writing—review and editing. All authors have read and agreed to the published version of the manuscript.

Funding: This research was funded by the National Science Centre, Poland (Decision No. 2021/41/B/ST8/00257).

Data Availability Statement: The data that support the findings of this study are openly available in the Mendeley data repository at http://doi.org/10.17632/7fbkf6y7gv.1 (accessed on 25 October 2021).

Conflicts of Interest: The authors declare no conflict of interest.

Appendix A

Gaussian Process for Regression

A Gaussian process (GP) defines a distribution $p(f)$ over the value of the function $f(\boldsymbol{x})$ at a finite set of points $\boldsymbol{x}_1, \ldots, \boldsymbol{x}_n$, wherein the collection of the random variables $f(\boldsymbol{x}_1), \ldots, f(\boldsymbol{x}_n)$ has a joint Gaussian distribution $p(f(\boldsymbol{x}_1), \ldots, f(\boldsymbol{x}_n))$ [20]. The GP is expressed as

$$f(\boldsymbol{x}) \sim GP(m(\boldsymbol{x}), k(\boldsymbol{x}, \boldsymbol{x}\prime)) \quad \text{(A1)}$$

where

$$m(\boldsymbol{x}) = \mathbb{E}[f(\boldsymbol{x})],\ k(\boldsymbol{x}, \boldsymbol{x}\prime) = \mathbb{E}[(f(\boldsymbol{x}) - m(\boldsymbol{x}))(f(\boldsymbol{x}\prime) - m(\boldsymbol{x}\prime))] \quad \text{(A2)}$$

are the mean and covariance functions, respectively. The covariance function (kernel) defines the inputs of the covariance matrix $\boldsymbol{K}$ of the joint Gaussian distribution:

$$K = \begin{bmatrix} k(\boldsymbol{x}_1, \boldsymbol{x}_1) & \ldots & k(\boldsymbol{x}_1, \boldsymbol{x}_n) \\ & \vdots & \\ k(\boldsymbol{x}_n, \boldsymbol{x}_1) & \ldots & k(\boldsymbol{x}_n, \boldsymbol{x}_n) \end{bmatrix} \quad \text{(A3)}$$

Because the joint Gaussian distribution is fully specified by the mean function m and covariance matrix $\boldsymbol{K}$, the function f can be randomly generated. The properties of function f are determined by the mean function $m(\boldsymbol{x})$ and kernel $k(\boldsymbol{x}, \boldsymbol{x}')$. However, the flexibility of the GP simplifies the calculation by using a zero mean function $m(\boldsymbol{x}) = 0$ [18,20]. Without the incorporation of any knowledge about observed inputs and outputs, the GP defines a prior distribution over the functions f. It can be transformed into a useful posterior distribution upon the application of the training data.

In the case of fatigue life prediction of structural materials analyzed herein, the function $f(x)$ is not accessible; only the noisy version y can be observed, $y = f(x) + \epsilon \sim \mathcal{N}\left(0, \sigma_y^2\right)$, where σ_y^2 is the variance. Thus, the diagonal of the covariance matrix $\boldsymbol{K}$ is increased as:

$$\boldsymbol{K}_y = \boldsymbol{K} + \sigma_y^2 \boldsymbol{I}, \tag{A4}$$

where $\boldsymbol{I}$ is an identity matrix of size n (the number of training points). The joint density of the observed outputs $\boldsymbol{y}$ and predicted function outputs $\boldsymbol{f}_*$

$$\begin{bmatrix} \boldsymbol{y} \\ \boldsymbol{f}_* \end{bmatrix} \sim \mathcal{N}\left(0, \begin{bmatrix} \boldsymbol{K}_y & \boldsymbol{K}_* \\ \boldsymbol{K}_*^T & \boldsymbol{K}_{**} \end{bmatrix}\right) \tag{A5}$$

involves the covariance matrix computed from the training $\boldsymbol{X}$ and test points $\boldsymbol{X}_*$, where $\boldsymbol{X}$ is a $d \times n$ matrix of the training inputs $\{x_i\}_{i=1}^n$, and $\boldsymbol{X}_*$ is the $d \times n_*$ matrix of the test inputs (n_* is the number of test points). Particularly, Equation (A5) includes the covariances $\boldsymbol{K}_* = k(\boldsymbol{X}, \boldsymbol{X}_*)$, $\boldsymbol{K}_y = k(\boldsymbol{X}, \boldsymbol{X}) + \sigma_y^2 \boldsymbol{I}$, $\boldsymbol{K}_{**} = k(\boldsymbol{X}_*, \boldsymbol{X}_*)$. The posterior predictive density was obtained [18] by conditioning the joint Gaussian distribution prior to the observation, as follows:

$$p(\boldsymbol{f}_* | \boldsymbol{X}_*, \boldsymbol{X}, \boldsymbol{y}) = \mathcal{N}(\boldsymbol{f}_* | \boldsymbol{\mu}_*, \boldsymbol{\Sigma}_*) \tag{A6}$$

where the searched median regression curve is represented by a vector of mean values $\boldsymbol{\mu}_*$ obtained for the test points $\boldsymbol{X}_*$ with the final expression as

$$\boldsymbol{\mu}_* = \boldsymbol{K}_*^T \boldsymbol{K}_y^{-1} \boldsymbol{y} \tag{A7}$$

and the variance of the function values for the test points $\boldsymbol{X}_*$ is included in the diagonal of the covariance matrix $\boldsymbol{\Sigma}_*$, denoted as

$$\boldsymbol{\Sigma}_* = \boldsymbol{K}_{**} - \boldsymbol{K}_*^T \boldsymbol{K}_y^{-1} \boldsymbol{K}_* \tag{A8}$$

The predicted outputs for the test points are specified by the above two equations, where the training data $(\boldsymbol{X}, \boldsymbol{y})$ serve as the parameters for regression. The covariance function k exclusively controls the predictive performance for a given set of training data. The kernel parameters (called hyperparameters) are obtained by maximizing the marginal likelihood, which is the maximum probability of $\boldsymbol{y}$ with noise for $p(\boldsymbol{y}|\boldsymbol{X}, \boldsymbol{\theta})$, where $\boldsymbol{\theta}$ is a vector of hyperparameters. The log marginal likelihood [20] is given by

$$log\ p(\boldsymbol{y}|\boldsymbol{X}, \boldsymbol{\theta}) = -\frac{1}{2} \boldsymbol{y}^T \boldsymbol{K}_y^{-1} \boldsymbol{y} - \frac{1}{2} \log|\boldsymbol{K}_y| - \frac{n}{2} \log 2\pi, \tag{A9}$$

where $|\boldsymbol{K}_y|$ is the determinant of the $\boldsymbol{K}_y$ matrix. The maximum marginal likelihood is found using a gradient-based optimizer. A quasi-Newton optimizer with a trust-region method was used in the analysis. The common stationary kernels were analyzed, as described in the next section. The training process was the same for all the covariance functions without the implementation of the cross-validation resampling procedure. The diagram of data flow for the fatigue life prediction based on the proposed GP fatigue model is presented in Figure A1. The diagram shows the exemplary data flow for four selected input parameters ($x = [\gamma_{ns}, \varepsilon_n, \tau_{ns}, \sigma_n]$ described in Section 4.3.1). The outputs are mean values of the log of fatigue lives for each test data point and its variance.

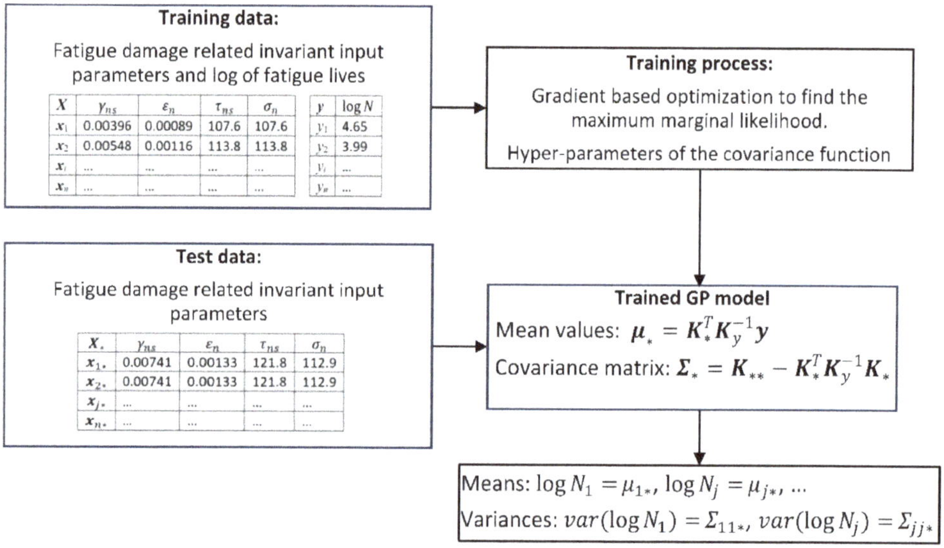

Figure A1. Diagram of data flow for the fatigue life prediction based on the proposed GP fatigue model.

Appendix B

Covariance Functions

The covariance function $k(x_p, x_q)$ determines the measure of similarity between outputs y_p and y_q based on the input points x_p, x_q locations. For stationary kernels (analyzed herein), only the distance between the point is used in the calculation, a radial basis function [18]. The smaller the distance between points, the higher the similarity between output values expected with a higher value of covariance. Each predictor included in the input vector x can be scaled by its length-scale parameter. If the value of the length-scale parameter (found by the maximum marginal likelihood optimization) approaches infinity, the corresponding dimension (predictor) can be ignored. This type of kernel is called an automatic relevance determination (ARD) covariance function. Such covariance functions were implemented to determine the crucial fatigue damage quantities for the tested CuZn37 brass. The distance r between points x_p, x_q is defined as

$$r = \sqrt{(x_p - x_q)^T M (x_p - x_q)} \tag{A10}$$

where $M = \text{diag}(l_i)^{-2}$ is a diagonal matrix of length-scale parameters $l_i = l_1, \ldots, l_d$ (each is assigned to a particular predictor). Four popular covariance functions were analyzed for fatigue life prediction, as presented below:

- Exponential (EX)

$$k(x_p, x_q) = \sigma_k^2 exp(-r) \tag{A11}$$

- Matern 3/2 (M3/2)

$$k(x_p, x_q) = \sigma_k^2 \left(1 + \sqrt{3}r\right) exp\left(-\sqrt{3}r\right) \tag{A12}$$

- Matern 5/2 (M5/2)

$$k(x_p, x_q) = \sigma_k^2\left(1 + \sqrt{5}r + \frac{5}{3}r^2\right)exp\left(-\sqrt{5}r\right) \quad \text{(A13)}$$

- Rational quadratic (RQ)

$$k(x_p, x_q) = \sigma_k^2\left(1 + \frac{r^2}{2\alpha}\right)^{-\alpha} \quad \text{(A14)}$$

- Squared exponential (SE)

$$k(x_p, x_q) = \sigma_k^2 exp\left(-\frac{1}{2}r^2\right). \quad \text{(A15)}$$

References

1. Stephens, R.I.; Fatemi, A.; Stephens, R.R.; Fuchs, H.O. *Metal Fatigue in Engineering*; Wiley: Hoboken, NJ, USA, 2000.
2. Wu, Y.L.; Zhu, S.P.; He, J.C.; Liao, D.; Wang, Q. Assessment of notch fatigue and size effect using stress field intensity approach. *Int. J. Fatigue* **2021**, *149*, 106279. [CrossRef]
3. Shamsaei, N.; Fatemi, A. Small fatigue crack growth under multiaxial stresses. *Int. J. Fatigue* **2014**, *58*, 126–135. [CrossRef]
4. Zhu, S.P.; Foletti, S.; Beretta, S. Probabilistic framework for multiaxial LCF assessment under material variability. *Int. J. Fatigue* **2017**, *103*, 371–385. [CrossRef]
5. Ai, Y.; Zhu, S.P.; Liao, D.; Correia, J.A.F.O.; Souto, C.; De Jesus, A.M.P.; Keshtegar, B. Probabilistic modeling of fatigue life distribution and size effect of components with random defects. *Int. J. Fatigue* **2019**, *126*, 165–173. [CrossRef]
6. Liao, D.; Zhu, S.P.; Keshtegar, B.; Qian, G.; Wang, Q. Probabilistic framework for fatigue life assessment of notched components under size effects. *Int. J. Mech. Sci.* **2020**, *181*, 105685. [CrossRef]
7. Karolczuk, A.; Macha, E. A review of critical plane orientations in multiaxial fatigue failure criteria of metallic materials. *Int. J. Fract.* **2005**, *134*, 267–304. [CrossRef]
8. Dantas, R.; Correia, J.; Lesiuk, G.; Rozumek, D.; Zhu, S.P.; de Jesus, A.; Susmel, L.; Berto, F. Evaluation of multiaxial high-cycle fatigue criteria under proportional loading for S355 steel. *Eng. Fail. Anal.* **2020**, *120*, 105037. [CrossRef]
9. Kenmeugne, B.; Soh Fotsing, B.D.; Anago, G.F.; Fogue, M.; Robert, J.L.; Kenne, J.P. On the evolution and comparison of multiaxial fatigue criteria. *Int. J. Eng. Technol.* **2012**, *4*, 37–46.
10. Arora, P.; Gupta, S.K.; Samal, M.K.; Chattopadhyay, J. Development of new critical plane model for assessment of fatigue life under multi-axial loading conditions. *Int. J. Fatigue* **2019**, *129*, 105209. [CrossRef]
11. Kalayci, C.B.; Karagoz, S.; Karakas, Ö. Soft computing methods for fatigue life estimation: A review of the current state and future trends. *Fatigue Fract. Eng. Mater. Struct.* **2020**, *43*, 2763–2785. [CrossRef]
12. Chen, J.; Liu, Y. Fatigue modeling using neural networks: A comprehensive review. *Fatigue Fract. Eng. Mater. Struct.* **2022**, *45*, 945–979. [CrossRef]
13. Zhang, C.; Wei, J.; Jing, H.; Fei, C.; Tang, W. Reliability-based low fatigue life analysis of turbine blisk with generalized regression extreme neural network method. *Materials* **2019**, *12*, 1545. [CrossRef]
14. An, D.; Kim, N.H.; Choi, J.H. Practical options for selecting data-driven or physics-based prognostics algorithms with reviews. *Reliab. Eng. Syst. Saf.* **2015**, *133*, 223–236. [CrossRef]
15. Kan, M.S.; Tan, A.C.C.; Mathew, J. A review on prognostic techniques for non-stationary and non-linear rotating systems. *Mech. Syst. Signal Process.* **2015**, *62*, 1–20. [CrossRef]
16. Zhou, Y.; Lu, Z. An enhanced Kriging surrogate modeling technique for high-dimensional problems. *Mech. Syst. Signal Process.* **2020**, *140*, 106687. [CrossRef]
17. *ASTM E739-10*; Standard Practice for Statistical Analysis of Linear or Linearized Stress-Life (S-N) and Strain-Life Fatigue Data. ASTM Int.: West Conshohocken, PA, USA, 2015.
18. Murphy, K.P. *Machine Learning: A Probabilistic Perspective*; The MIT Press: Cambridge, MA, USA, 2012; ISBN 9780262018029.
19. Wan, H.P.; Ren, W.X.; Todd, M.D. An efficient metamodeling approach for uncertainty quantification of complex systems with arbitrary parameter probability distributions. *Int. J. Numer. Methods Eng.* **2017**, *109*, 739–760. [CrossRef]
20. Rasmussen, C.E.; Williams, C.K.I. *Gaussian Process for Machine Learning*; The MIT Press: Cambridge, MA, USA, 2006; ISBN 026218253X.
21. Huchet, Q.; Mattrand, C.; Beaurepaire, P.; Relun, N.; Gayton, N. AK-DA: An efficient method for the fatigue assessment of wind turbine structures. *Wind Energy* **2019**, *22*, 638–652. [CrossRef]
22. Aye, S.A.; Heyns, P.S. An integrated Gaussian process regression for prediction of remaining useful life of slow speed bearings based on acoustic emission. *Mech. Syst. Signal Process.* **2017**, *84*, 485–498. [CrossRef]
23. Hao, H.; Ye, D.; Chen, C. Strain ratio effects on low-cycle fatigue behavior and deformation microstructure of 2124-T851 aluminum alloy. *Mater. Sci. Eng. A* **2014**, *605*, 151–159. [CrossRef]

24. Mohanty, S.; Chattopadhyay, A.; Peralta, P.; Das, S.; Willhauck, C. Fatigue life prediction using multivariate gaussian process. In Proceedings of the 49th AIAA/ASME/ASCE/AHS/ASC Structures, Structural Dynamics, and Materials Conference, Schaumburg, IL, USA, 7–10 April 2008; pp. 1–14.
25. Mohanty, S.; Chattopadhyay, A.; Peralta, P.; Das, S. Bayesian Statistic Based Multivariate Gaussian Process Approach for Offline/Online Fatigue Crack Growth Prediction. *Exp. Mech.* **2011**, *51*, 833–843. [CrossRef]
26. Hirvoas, A.; Prieur, C.; Arnaud, E.; Caleyron, F.; Munoz Zuniga, M. Quantification and reduction of uncertainties in a wind turbine numerical model based on a global sensitivity analysis and a recursive Bayesian inference approach. *Int. J. Numer. Methods Eng.* **2021**, *122*, 2528–2544. [CrossRef]
27. Słoński, M. Bayesian neural networks and Gaussian processes in identification of concrete properties. *Comput. Assist. Mech. Eng. Sci.* **2011**, *18*, 291–302.
28. Hu, D.; Su, X.; Liu, X.; Mao, J.; Shan, X.; Wang, R. Bayesian-based probabilistic fatigue crack growth evaluation combined with machine-learning-assisted GPR. *Eng. Fract. Mech.* **2020**, *229*, 106933. [CrossRef]
29. Ling, Y.; Mahadevan, S. Integration of structural health monitoring and fatigue damage prognosis. *Mech. Syst. Signal Process.* **2012**, *28*, 89–104. [CrossRef]
30. Farid, M. Data-driven method for real-time prediction and uncertainty quantification of fatigue failure under stochastic loading using artificial neural networks and Gaussian process regression. *Int. J. Fatigue* **2021**, *155*, 106415. [CrossRef]
31. Flah, M.; Nunez, I.; Ben Chaabene, W.; Nehdi, M.L. Machine Learning Algorithms in Civil Structural Health Monitoring: A Systematic Review. *Arch. Comput. Methods Eng.* **2020**, *28*, 2621–2643. [CrossRef]
32. Karolczuk, A.; Słoński, M. Application of the Gaussian process for fatigue life prediction under multiaxial loading. *Mech. Syst. Signal Process.* **2022**, *167*, 108599. [CrossRef]
33. Karolczuk, A.; Papuga, J.; Palin-luc, T. Progress in fatigue life calculation by implementing life-dependent material parameters in multiaxial fatigue criteria. *Int. J. Fatigue* **2020**, *134*, 105509. [CrossRef]
34. Pejkowski, Ł.; Seyda, J. Fatigue of four metallic materials under asynchronous loadings: Small cracks observation and fatigue life prediction. *Int. J. Fatigue* **2021**, *142*, 105904. [CrossRef]
35. Niu, X.P.; Wang, R.Z.; Liao, D.; Zhu, S.P.; Zhang, X.C.; Keshtegar, B. Probabilistic modeling of uncertainties in fatigue reliability analysis of turbine bladed disks. *Int. J. Fatigue* **2021**, *142*, 105912. [CrossRef]
36. Karolczuk, A.; Kluger, K.; Palin-luc, T. Fatigue failure probability estimation of the 7075-T651 aluminum alloy under multiaxial loading based on the life-dependent material parameters concept. *Int. J. Fatigue* **2021**, *147*, 106174. [CrossRef]
37. Karolczuk, A.; Kurek, M. Fatigue life uncertainty prediction using the Monte Carlo and Latin hypercube sampling techniques under uniaxial and multiaxial cyclic loading. *Int. J. Fatigue* **2022**, *160*, 106867. [CrossRef]
38. Macha, E.; Sonsino, C.M. Energy criteria of multiaxial fatigue failure. *Fatigue Fract. Eng. Mater. Struct.* **1999**, *22*, 1053–1070. [CrossRef]
39. Skibicki, D. *Phenomena and Computational Models of Non-Proportional Fatigue of Materials*; Springer: Berlin/Heidelberg, Germany, 2014; Volume 18, ISBN 9783319015644.
40. Gough, H.J. Engineering steels under combined cyclic and static stresses. *J. Appl. Mech. ASME* **1950**, *17*, 113–125. [CrossRef]
41. Gough, H.J.; Clenshaw, W.J.; Pollard, H.V. *Some Experiments on the Resistance of Metals to Fatigue under Combined Stresses*; His Majesty's Stationery Office: London, UK, 1951.
42. Pejkowski, Ł.; Skibicki, D.; Kurek, M. Generalization of Gough and Pollard Criterion for the Finite Fatigue Life Regime in Proportional Loading Conditions. *J. Test. Eval.* **2019**, *47*, 1023–1046. [CrossRef]
43. Sines, G. *Failure of Materials under Combined Repeated Stresses with Superimposed Static Stresses*; University of California: Los Angeles, CA, USA, 1955.
44. Crossland, B. Effect of large hydrostatic pressures on the torsional fatigue strength of an alloy steel. In Proceedings of the International Conference on Fatigue of Metals, London, UK, 10–14 September 1956; Institute of Mechanical Engineers: London, UK, 1956; pp. 138–149.
45. Papuga, J. A survey on evaluating the fatigue limit under multiaxial loading. *Int. J. Fatigue* **2011**, *33*, 153–165. [CrossRef]
46. Polák, J.; Man, J. Experimental evidence and physical models of fatigue crack initiation. *Int. J. Fatigue* **2016**, *91*, 294–303. [CrossRef]
47. Mughrabi, H. Microstructural mechanisms of cyclic deformation, fatigue crack initiation and early crack growth. *Philos. Trans. R. Soc. A Math. Phys. Eng. Sci.* **2015**, *373*, 20140132. [CrossRef]
48. Findley, W.N. A theory for the effect of mean stress on fatigue of metals under combined torsion and axial load or bending. *Dep. Army Proj.* **1958**, *81*, 301–305. [CrossRef]
49. Brown, M.W.; Miller, K.J. A Theory for Fatigue Failure under Multiaxial Stress-Strain Conditions. *Proc. Inst. Mech. Eng.* **1973**, *187*, 745–755. [CrossRef]
50. Kandil, F.A.; Brown, M.W.; Miller, K.J. *Biaxial Low-Cycle Fatigue Failure of 316 Stainless Steel at Elevated Temperatures*; Metals Society: London, UK, 1982; pp. 203–210.
51. Fatemi, A.; Socie, D.F. A critical plane approach to multiaxial fatigue damage including out-of-phase loading. *Fatigue Fract. Eng. Mater. Struct.* **1988**, *11*, 149–165. [CrossRef]
52. Carpinteri, A.; Spagnoli, A.; Vantadori, S. Multiaxial fatigue assessment using a simplified critical plane-based criterion. *Int. J. Fatigue* **2011**, *33*, 969–976. [CrossRef]
53. Papuga, J.; Růžička, M. Two new multiaxial criteria for high cycle fatigue computation. *Int. J. Fatigue* **2008**, *30*, 58–66. [CrossRef]

54. Papuga, J.; Halama, R. Mean stress effect in multiaxial fatigue limit criteria. *Arch. Appl. Mech.* **2019**, *89*, 823–834. [CrossRef]
55. Glinka, G.; Shen, G.; Plumtree, A. A multiaxial fatigue strain energy density parameter related to the critical fracture plane. *Fatigue Fract. Eng. Mater. Struct.* **1995**, *18*, 37–46. [CrossRef]
56. Pan, W.F.; Hung, C.Y.; Chen, L.L. Fatigue life estimation under multiaxial loadings. *Int. J. Fatigue* **1999**, *21*, 3–10. [CrossRef]
57. Ince, A.; Glinka, G. A generalized fatigue damage parameter for multiaxial fatigue life prediction under proportional and non-proportional loadings. *Int. J. Fatigue* **2014**, *62*, 34–41. [CrossRef]
58. Yu, Z.Y.; Zhu, S.P.; Liu, Q.; Liu, Y. Multiaxial fatigue damage parameter and life prediction without any additional material constants. *Materials* **2017**, *10*, 923. [CrossRef]
59. Gates, N.R.; Fatemi, A. On the consideration of normal and shear stress interaction in multiaxial fatigue damage analysis. *Int. J. Fatigue* **2017**, *100*, 322–336. [CrossRef]
60. Gates, N.; Fatemi, A. Friction and roughness induced closure effects on shear-mode crack growth and branching mechanisms. *Int. J. Fatigue* **2016**, *92*, 442–458. [CrossRef]
61. Papadopoulos, I.V.; Davoli, P.; Gorla, C.; Filippini, M.; Bernasconi, A. A comparative study of multiaxial high-cycle fatigue criteria for metals. *Int. J. Fatigue* **1997**, *19*, 219–235. [CrossRef]
62. *ASTM E2207-02*; Standard Practice for Strain-Controlled Axial-Torsional Fatigue Testing with Thin-Walled Tubular Specimens. ASTM Int.: West Conshohocken, PA, USA, 2002.
63. Karolczuk, A.; Skibicki, D.; Pejkowski, Ł. Evaluation of the Fatemi-Socie damage parameter for the fatigue life calculation with application of the Chaboche plasticity model. *Fatigue Fract. Eng. Mater. Struct.* **2019**, *42*, 197–208. [CrossRef]
64. Skibicki, D.; Pejkowski, Ł. Low-cycle multiaxial fatigue behaviour and fatigue life prediction for CuZn37 brass using the stress-strain models. *Int. J. Fatigue* **2017**, *102*, 18–36. [CrossRef]
65. Beretta, S.; Foletti, S.; Rusconi, E.; Riva, A.; Socie, D.F. A log-normal format for failure probability under LCF: Concept, validation and definition of design curve. *Int. J. Fatigue* **2016**, *82*, 2–11. [CrossRef]
66. Papadopoulos, I.V. Long life fatigue under multiaxial loading. *Int. J. Fatigue* **2001**, *23*, 839–849. [CrossRef]
67. Schijve, J. Statistical distribution functions and fatigue of structures. *Int. J. Fatigue* **2005**, *27*, 1031–1039. [CrossRef]
68. Bastenaire, F.A.A. New method for the statistical evaluation of constant stress amplitude fatigue-test results. *ASTM Spec. Tech. Publ.* **1971**, 3–28. [CrossRef]
69. Wang, Y.; Faruq, N.Z.; Susmel, L. Evaluation of different techniques in estimating orientation of crack initiation planes and fatigue lifetime under complex multiaxial loading paths. *Int. J. Fatigue* **2017**, *100*, 521–529. [CrossRef]
70. Anes, V.; Reis, L.; Li, B.; Freitas, M.; Sonsino, C.M. Minimum Circumscribed Ellipse (MCE) and Stress Scale Factor (SSF) criteria for multiaxial fatigue life assessment. *Theor. Appl. Fract. Mech.* **2014**, *73*, 109–119. [CrossRef]
71. Xu, S.; Zhu, S.P.; Hao, Y.Z.; Liao, D. Critical plane–based multiaxial fatigue life prediction of turbine disk alloys by refining normal stress sensitivity. *J. Strain Anal. Eng. Des.* **2018**, *53*, 719–729. [CrossRef]
72. Skibicki, D.; Pejkowski, Ł. Integral fatigue criteria evaluation for life estimation under uniaxial combined proportional and non-proportional loadings. *J. Theor. Appl. Mech.* **2012**, *50*, 1073–1086.
73. Mršnik, M.; Slavič, J.; Boltežar, M. Multiaxial vibration fatigue—A theoretical and experimental comparison. *Mech. Syst. Signal Process.* **2016**, *76–77*, 409–423. [CrossRef]
74. Karolczuk, A.; Kluger, K.; Łagoda, T. A correction in the algorithm of fatigue life calculation based on the critical plane approach. *Int. J. Fatigue* **2016**, *83*, 174–183. [CrossRef]
75. Fatemi, A.; Kurath, P. Multiaxial fatigue life predictions under the influence of mean-stresses. *J. Eng. Mater. Technol. ASME* **1988**, *110*, 380–388. [CrossRef]
76. Pejkowski, Ł.; Karuskevich, M.; Maslak, T. Extrusion/intrusion structure as a fatigue indicator for uniaxial and multiaxial loading. *Fatigue Fract. Eng. Mater. Struct.* **2019**, *42*, 2315–2324. [CrossRef]
77. Sakane, M.; Itoh, T. Microstructural study of multiaxial low cycle fatigue. *Frat. Integritá Strutt.* **2015**, *9*, 319–334. [CrossRef]
78. Skibicki, D. Multiaxial fatigue life and strength criteria for non-proportional loading. *Mater. Test.* **2006**, *48*, 99–102. [CrossRef]
79. Itoh, T.; Kameoka, M.; Obataya, Y. A new model for describing a stable cyclic stress-strain relationship under non-proportional loading based on activation state of slip systems. *Fatigue Fract. Eng. Mater. Struct.* **2004**, *27*, 957–966. [CrossRef]
80. Kang, G.; Gao, Q.; Yang, X. Uniaxial and non-proportionally multiaxial ratcheting of SS304 stainless steel at room temperature: Experiments and simulations. *Int. J. Non Linear. Mech.* **2004**, *39*, 843–857. [CrossRef]
81. Meggiolaro, M.A.; De Castro, J.T.P. Prediction of non-proportionality factors of multiaxial histories using the Moment of Inertia method. *Int. J. Fatigue* **2014**, *61*, 151–159. [CrossRef]
82. Skibicki, D.; Sempruch, J. Use of a load non-proportionality measure in fatigue under out-of-phase combined bending and torsion. *Fatigue Fract. Eng. Mater. Struct.* **2004**, *27*, 369–377. [CrossRef]
83. Meggiolaro, M.A.; De Castro, J.T.P. An improved multiaxial rainflow algorithm for non-proportional stress or strain histories—Part I: Enclosing surface methods. *Int. J. Fatigue* **2012**, *42*, 217–226. [CrossRef]
84. Papuga, J.; Cízová, E.; Karolczuk, A. Validating the methods to process the stress path in multiaxial high-cycle fatigue criteria. *Materials* **2021**, *14*, 206. [CrossRef]

materials

MDPI

Article

The Influence of the Type of Fibers on the Reduction of the Threshold Effect in the Transition Zone of a Railway Track

Włodzimierz Idczak *, Tomasz Lewandrowski, Dominik Pokropski, Grzegorz Rogojsz and Tomasz Rudnicki

Faculty of Civil Engineering and Geodesy, Military University of Technology, 2 Gen. Sylwestra Kaliskiego Str., 00-908 Warsaw, Poland
* Correspondence: wlodzimierz.idczak@wat.edu.pl

Abstract: The presented article concentrates on the influence of various concrete additives in the form of fibers on the mechanical parameters of concrete so as to obtain the effect of gradual changes in these parameters, which is very important in the transition zone of the railway track. Steel, polymer and glass fibers, as well as concrete without additives, were accepted for the study. The effect of additives on the consistency of the mixture, compressive strength, frost resistance and elastic modulus was studied. The research concerned concrete samples and models of elements of the ballastless railway surface, i.e., track slab and concrete block supports. The track slab model was made of concrete without additives, while the models of supports were made both without and with additives. The studies were carried out in laboratory conditions. As a result, the tested concrete samples with various additives were ranked so that they could be used as a material for elements of the railway surface in the transition zones of engineering facilities on railway roads, which is important from the point of view of reducing the threshold effect occurring in these zones. Detailed laboratory tests were presented, the results of these studies were discussed, and final conclusions were drawn regarding the technology of materials and the methodology of constructing the transition zones of the railway surface in order to avoid or at least reduce the threshold effect existing in such zones.

Keywords: concrete additives; concrete fibers; concrete strength tests; threshold effect

Citation: Idczak, W.; Lewandrowski, T.; Pokropski, D.; Rogojsz, G.; Rudnicki, T. The Influence of the Type of Fibers on the Reduction of the Threshold Effect in the Transition Zone of a Railway Track. *Materials* **2022**, *15*, 5730. https://doi.org/10.3390/ma15165730

Academic Editor: Andreas Lampropoulos

Received: 1 July 2022
Accepted: 18 August 2022
Published: 19 August 2022

1. Introduction

In previous practice, concrete additives in the form of mineral ingredients or chemical admixtures were used for two reasons: (1) to obtain concretes with increased strength parameters, and (2) to use waste raw materials and to increase the recycling rate. This article describes the use of concrete additives in the form of various fibers in order to obtain a gentle change of the strength parameters of fragments of the railway surface on the section of the transition zone in order to eliminate, or at least reduce, the impact of the threshold effect on the dynamic interactions at the interface of the rail vehicle with the railway track.

1.1. A Brief Overview of Previous Studies

With the development of construction, there is an increase in demand for concretes with high strength parameters. Due to the occurrence of concrete shrinkage, it is not possible to increase the cement content in it indefinitely. For this reason, a number of mineral additives and chemical admixtures have been used for concrete for decades [1–4]. One of them is diffuse reinforcement in the form of fibers. Currently, such solutions are used in ceilings, tunnels, foundations, road construction and even prefabricated elements [5–8]. The main task of diffuse reinforcement is to prevent the formation of shrinkage microcracks in concrete. Moreover, fibers for concrete have a beneficial effect on mechanical and operational parameters after bonding the concrete mix. They may improve the compressive strength, as well as improve the frost resistance of concrete. Some modern techniques for concrete reinforcement with polymers were widely described in [9]. In addition, the use of additives in the form of fibers for the production of concrete has an ecological aspect, because in this case, waste materials, such as used glass or plastics, become additives. The use of used glass or plastics for the production of additives for concrete can improve the recycling rate of these materials. For example, glass recycling in Poland is about 62% and is several percent lower than the average for the European Union. In many countries, this indicator reaches up to 95%. Increasing the use of concrete additives makes it possible to increase this indicator so that as little waste as possible goes to landfills [10]. Another very interesting study, concerning the influence of certain recycled aggregates added in place of sand on the mechanical properties of concrete, is presented in [11]. Extending the life cycle of concrete used in road construction can be achieved, for example, by 100% recycling of the concrete surface even after 80 years or by using cements with a low carbon footprint. A very important element in improving the durability of the road surface is the use of fibers and chemical admixtures that increase the plasticity of the mixture and shape the structure of aeration of hardened concrete [12].

1.2. New Applications

The use of various concrete additives to obtain materials with different, slightly changing strength parameters has not been the subject of research described in the available literature. This type of research was initiated in the work presented in [13]. This article is a continuation of the research presented there, where a simple computational theoretical model of the interaction of a rail vehicle with a railway track, which made it possible to study dynamic phenomena arising in the zones of change of railway surface technology, was described. In the literature on the subject for calculations, a model of the railway surface is adopted, in which the rail is based on the elastic Winkler substrate characterized by the modulus of elasticity of the substrate. The validity of the adopted model was confirmed by previous research and analytical work, among others, presented in [14]. The validity of this approach has also been proven, inter alia, in [15], where a number of variants of rail, including four ways of mapping a vehicle in the form of a stream of concentrated forces, a stream of concentrated masses and streams of one- and two-mass oscillators, were analyzed. The same representations of a moving-rail vehicle were also used in [16–18], each time giving correct, real results. The theoretical model described in [13] was experimentally verified using laser scanning technology on active sections of the railway line. The advantages of measurements performed in this technology include high accuracy and automation, as well as the speed of measurements and the lack of the need to destroy the tested object or exclude it from operation. The main element of the measuring set included the laser scanner of the scanCONTROL LLT2610-50 series [19]. Thanks to positive experimental verification of the computational theoretical model, further theoretical analyses were possible for any/variable material technologies of surfaces in transition zones where the threshold phenomenon occurs. The threshold effect has a negative impact not only on the railway surface, but on the object that is exposed to excessive loads and vibrations as well [20–22]. The negative effects of the threshold effect were reduced in [13] by modifications of the railway ballast surface. In this article, concerning experimental research, an attempt is made to modify the ballastless section of the railway surface. Thanks to this, after applying the obtained set of materials in the elements of the railway surface in the transition zones: in front of and behind the railway engineering infrastructure facility, a significant reduction in the adverse effects of the threshold effect that occurs in these zones will be achieved. Thus, the next points of the article describe research on the mechanical properties of the material samples themselves, as well as tests of laboratory models of specific elements of the railway track, used in the transition zones of the railway line, made of previously tested materials.

2. Materials and Methods

2.1. Materials

The composition of the concrete mix intended for the track slab has been designed on the basis of the method of three equations in accordance with the requirements set out in the technical documents [23–26]. In the further part of the article, cement concrete intended for the track plate is marked with the symbol Z1.

Due to the fact that block supports are prefabricated elements and the design of the concrete mix is a trade secret, the components of concrete intended for block supports were selected on the basis of information obtained from the manufacturer of the ÖBB-PORR system and analogies to the requirements set out in "the Technical Conditions for the Execution and Acceptance of Prestressed Concrete Sleepers and Turnouts Id-101" [27]. In order to analyze the strength parameters in concrete intended for block supports, an additive in the form of fibers was used. Concretes in the further part of the article will be marked as:

Z2—concrete with the addition of steel fibers, Z3—concrete with the addition of polymer fibers, Z4—concrete with the addition of glass fibers, Z5—reference concrete, without the addition of fibers.

For concrete mixes, fine aggregate 0/2 from the Żabiny mine (Morąg, Poland) and granite coarse aggregate with fractions 2/8 and 8/16 from the Graniczna mine (Strzegom, Poland) were used. The concrete mix intended for the Z1 track slab was made using CEM I 42.5 N/NA cement from the WARTA company (Trębaczew, Poland), while CEM 42.5 R cement from the CEMEX company (Warsaw, Poland) was used to make the concrete mix intended for Z2–Z5 block supports. The properties of the cements used are shown in Table 1. Design assumptions for concrete mixtures are listed in Table 2 and laboratory prescriptions in Tables 3 and 4.

Table 1. Properties of cements.

Parameter	Unit	CEM I 42.5 N/NA WARTA	CEM I 42.5 R CEMEX
Compressive strength			
after 2 days	MPa	17.9	25.6
after 28 days	MPa	50.5	54.3
Bending strength			
after 2 days	MPa	3.7	-
after 28 days	MPa	8.3	-
Setting time:			
beginning	min	200	187
the end	min	280	248
Specific surface	cm^2/g	3207	3746
Cement shrinkage	mm/m	0.46	-
Loss of roasting	%	3.80	3.24
Chemical composition:			
CaO	%	63.69	-
MgO	%	1.07	-
SiO_2	%	20.19	-
Al_2O_3	%	4.49	-
Fe_2O_3	%	2.78	-
SO_3	%	2.36	2.98
Na_2Oeq	%	0.54	0.60
Cl	%	0.032	0.079

Table 2. Assumptions for concrete mixes.

Parameter	Concrete for Track Slab	Concrete for Block Supports
Compressive strength class	C35/45	C50/60
Exposure class to carbonation corrosion	XC4	XC4
Frost aggression exposure class	XF1	XF1
Degree of frost resistance	F150	F125
Required cement grade	CEM 42.5	CEM 42.5
Maximum cement content [kg]	400	450
Required consistency	S3	S2
Maximum absorbability of concrete	5%	4%
Water-cement indicator w/c	<0.5	<0.45

Table 3. Concrete mix design for track slab (Z1).

Ingredient	Content [kg]
Cement CEM I 42.5 N/NA from WARTA company	390
Fine aggregate Sand 0/2 from Żabiny company	517
Coarse aggregate 1 Granite 2/8 from Graniczna company	497
Coarse aggregate 2 Granit 8/16 from Graniczna company	919
Water	148
Plasticizer Version of Viscocrete	3.1
Aerator Sica AerPro 3	0.6

Table 4. Design of concrete mix for block supports.

Ingredient	Content [kg]
Cement CEM I 42.5 R from CEMEX company	420
Fine aggregate Sand 0/2 from Żabiny company	486
Coarse aggregate 1 Granite 2/8 from Graniczna company	523
Coarse aggregate 2 Granit 8/16 from Graniczna company	860
Water	155
Plasticizer Version of Viscocrete	3.3
Aerator Sica AerPro 3	0.7

The dosage of fibers in individual mixtures meets the requirements specified by the manufacturer and the standard provisions [28,29].

In the concrete mix intended for block supports Z2, steel fibers for concrete were used, from the Siatpol company (Majdan Stary, Poland) with a length of 50 mm in the amount of 25 kg/m^3 of concrete mix. The concrete mix intended for Z3 block supports uses Polyex Duro polymer fibers from the Astra company (Straszyn, Poland) with a length of 25 mm, in the amount of 4.5 kg/m^3 of concrete mix. The Z4 blend uses CemFil Hp Macro glass fibers from the Serra-Ciments company (Barcelona, Spain) with a length of 36 mm in the amount of 1 kg/m^3 of concrete mix. The dosage of the fibers was based on the manufacturer's recommendations and the standard provisions [28,30]. The parameters of fibers used are summarized in Table 5 and shown in Figure 1.

Table 5. Parameters of fibers.

Parameter	Steel Fibers	Polymer Fibers	Glass Fibers
Diameter	1.0 mm ± 10%	0.6 mm	0.01 mm
Length	50 mm ± 10%	25 mm ± 10%	36 mm ± 10%
Dosage	25–35 kg/m^3	3–9 kg/m^3	0.5–2 kg/m^3
Tensile strength	1100 MPa	550–650 MPa	2500 MPa

(a) **(b)** **(c)**

Figure 1. Fibers used: (**a**) steel, (**b**) polymer, (**c**) glass.

2.2. Research Methods

The first study that was carried out was the study of the consistency of the concrete mix. The consistency of the mixture was determined by the Abrams cone fall method. The consistency test was carried out in accordance with PN-EN 12350-2:2011 Concrete mix tests—part 2: Consistency testing by the cone fall method [31].

Another test was the compressive strength test of cement concrete samples. This test was performed on cubic samples with dimensions of 150 mm × 150 mm × 150 mm in accordance with the provisions of PN-EN 12390-3:2019 Concrete tests—Part 3: Compressive strength of samples for testing [32]. Due to the possibility of random results, the number of samples to be tested was determined in accordance with PN-EN 206:2014 Concrete—Requirements, properties, production, and conformity [23]. The test was carried out with the use of the FORM TEST MEGA 6 3000-150 testing machine (Figure 2). Concrete samples after the compressive strength test are presented in Figure 3.

Figure 2. Sample during compressive strength test.

Figure 3. Concrete samples after compressive strength test.

The next study aimed at determining the parameters of the designed cement concretes was the study of the frost resistance of concrete. This test was carried out by the usual method in order to verify the degree of frost resistance F assumed at the design stage of the mixture. However, this degree in the assumptions corresponds to the N index, which is equal to the number of expected years of use of the structure. The usual method makes it possible to take into account the degree of internal destruction of concrete, characterized by the decreased strength of the sample, as well as the external destruction, determined by the loss of mass of the sample. The frost resistance test was carried out in accordance with PN-88/B-06250:1988 Ordinary concrete [33]. The test was carried out with the use of the TOROPOL K-15 testing machine. Samples prepared for frost resistance testing in this device are shown in Figure 4.

Figure 4. Samples prepared for frost resistance testing.

In order to determine the physical parameters of the designed cement concretes, a study of the elastic modulus was carried out. Cylindrical samples with a diameter of 150 mm and a height of 300 mm were prepared for the study. The study of elastic modulus was carried out using two methods. The first method consisted in determining the modulus of elasticity using the resonant method. In this method, the modulus of elasticity is determined by the propagation of the waves in the sample. Vibrations are caused by hitting a ball of a properly selected diameter on the base of the cylinder. Based on the measured wave, the device determines the frequency of vibrations and the modulus of elasticity of the tested material. This test is a non-destructive test of the element under study. The test was carried out with the use of the JAMES INSTRUMENT V-E-400 testing machine (Figure 5). In the second method, tests of the elastic modulus were carried out in accordance with PN-EN 12390-13:2014 Concrete tests—Part 13: Determination of the secant modulus of elasticity at compression [34]. The test was carried out with the use of the FORM TEST MEGA 6 3000-150 testing machine (Figure 6).

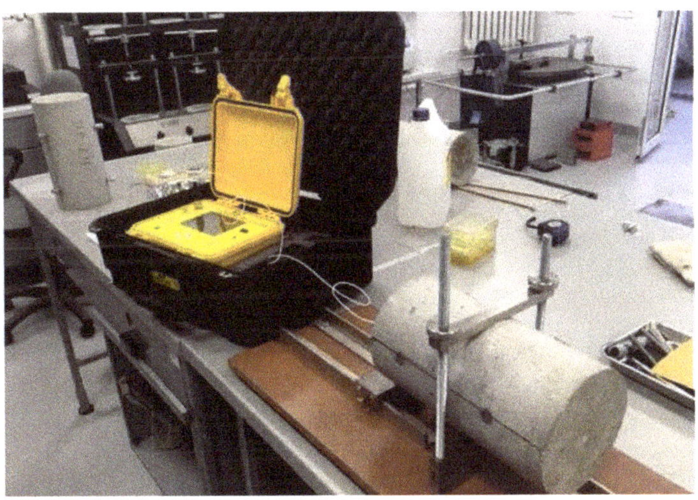

Figure 5. Resonance testing of the modulus of elasticity.

Figure 6. Testing of the secant modulus of elasticity at compression.

The main part of the experimental research was to determine the impact of the stiffness and strength of block supports on the strength of the entire ballastless surface, and thus on the strength of the track plate made of cement concrete. According to the adopted model, there are tensions of concrete in the track plate. At the same time, under the block supports, the forces from the wheeled vehicle cause the concrete to stretch downwards, and in the middle of the length of the slab the phenomenon of lifting the slab is triggered. The plate load model adopted for the cross-sectional test is shown in Figure 7.

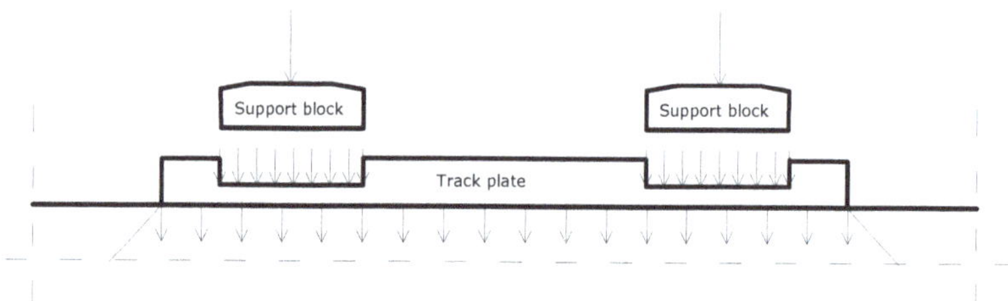

Figure 7. Load distribution model adopted for the test.

Due to the research possibilities, in order to check the tensile strength when bending the track plate with embedded block supports, a laboratory sample constituting 30% of the actual dimensions was prepared. The dimensions of the sample on which the tests were carried out were 100 mm × 150 mm × 750 mm and were adapted to the forms equipped by the laboratory. Figure 8 shows a diagram of the test samples, and Figure 9 shows the sample.

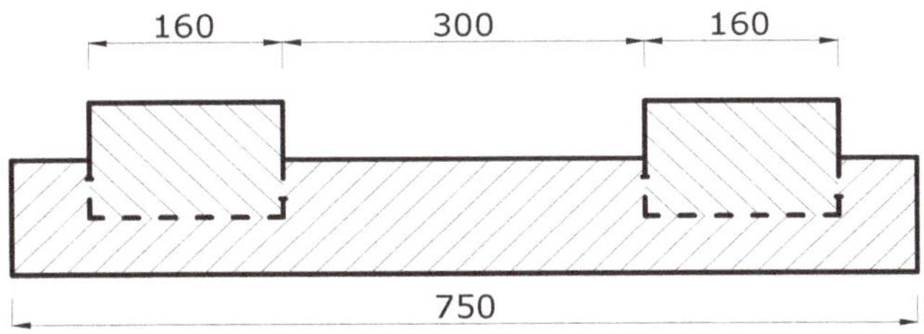

Figure 8. Cross-section of the laboratory sample (dimensions are described in mm).

Figure 9. Laboratory sample prepared for testing.

The tests (Figure 10) were carried out for all cement concretes Z2–Z5 intended for use in block supports. The load was applied in the middle of the block supports so as to map the forces occurring in the structure. The support points were taken at a distance of 5 cm from the edge of the sample. The adoption of such a spacing of supports was aimed at recreating the variant of the destruction of the substructure during the operation of the railway line.

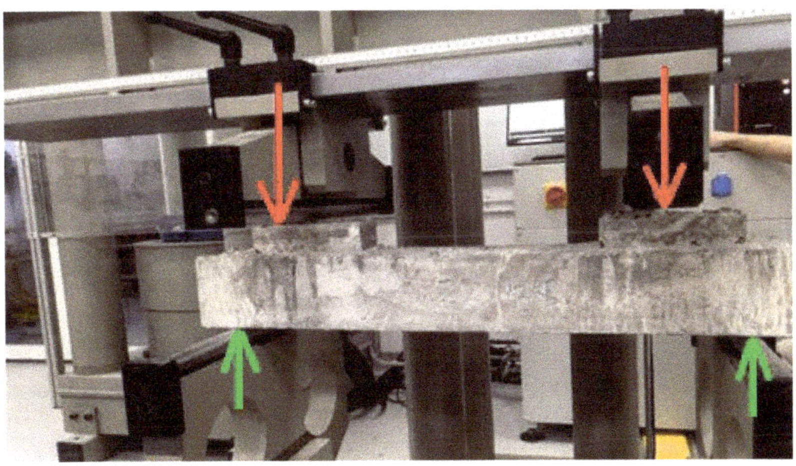

Figure 10. Sample in the test machine (in a model: red arrows—forces generated by the vehicle wheel, green arrows—reaction forces generated by the machine support).

This destruction consists in settling or washing the foundation from the central part of the support. The stress value of the samples is based on the following formula.

$$\sigma_g = \frac{P \cdot z}{2 \cdot W_g} \quad (1)$$

where:

P—force acting on the sample (kN), z—distance of force from the support—0.05 m, *Wg*—index of the bending strength of the rectangular cross-section relative to the vertical axis.

Another test aimed at determining the strength of block supports on the strength of the entire surface was to check the strength of the longitudinal section. The purpose of this test was to check the load capacity in the direction along the track, derived from the axle load of the bogie of the AEG12X locomotive, with a spacing of 2.60 m. The axle load of the locomotive is 210 kN. The plate load model adopted for longitudinal section testing is shown in Figure 11.

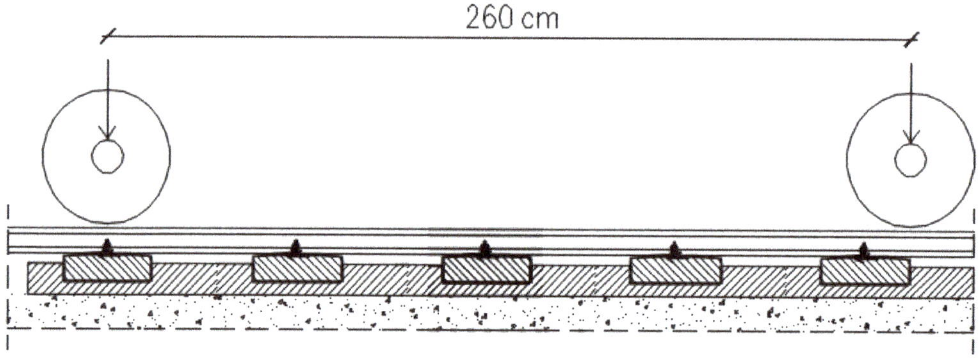

Figure 11. Scheme of loading the railway surface with a locomotive.

In order to check the tensile strength when bending along the surface, samples measuring 10 × 15 × 100 cm were prepared. The construction of the surface was made on a scale of 0.3. It is therefore the same scale factor as in the case of cross-sectional testing. In addition to the limited availability of large molds for making samples, the choice of scale factor in this study was dictated by the capabilities of the strength testing machine. Additionally to the limited availability of large molds for making samples, the choice of scale factor in this study was dictated by the capabilities of the strength testing machine. Figure 12 shows the designed test sample with dimensions in cm. In addition, Figure 13 shows the prepared sample for testing.

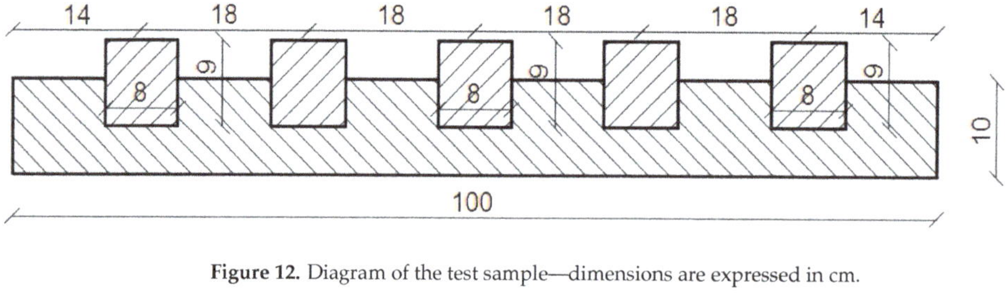

Figure 12. Diagram of the test sample—dimensions are expressed in cm.

Figure 13. Prepared sample for bending strength testing.

A sample after the bending strength testing is shown in Figure 14 (test machine is shown in Figure 10).

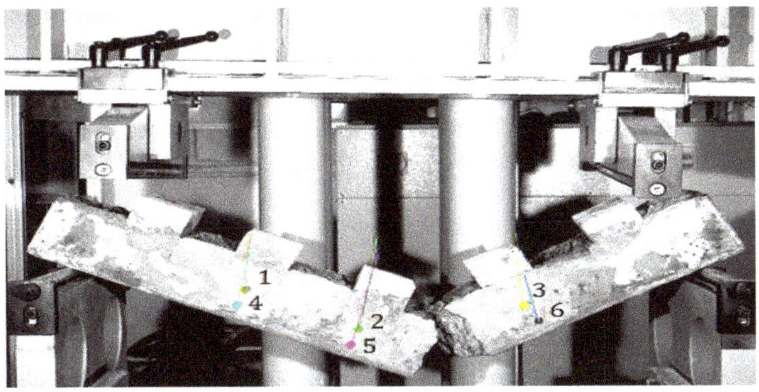

Figure 14. Sample after bending strength testing—dots were used to register the movement of the sample with the PHANTOM MICRO LC310 camera—not described in this article.

The research was carried out for four types of block supports. The load was applied, as in the previous test, in the middle of the block supports. Support points are adopted at a distance of 5 cm from the edge. The stress values in the samples were determined in the same way as in the previous test, i.e., according to Equation (1).

3. Results

3.1. Consistency of Concrete Mix

The results of measuring the consistency of the concrete mix are presented in Table 6.

Table 6. Results of the measurement of cone fall.

Concrete Type	Cone Fall [mm]	Consistency Class
Z1	140	S3
Z2	80	S2
Z3	70	S2
Z4	60	S2
Z5	80	S2

The greatest cone fallout was observed in the case of concrete without the addition of fibers and with steel fibers. The smallest for concrete with the addition of glass fibers. The reason for this is the effect of fibers on the workability and consistency of the mixture. Fine polymer and glass fibers propagated in the mixture, improving its stability. Steel fibers, due to their greater rigidity and larger size, did not provide this property. Concrete without the addition of fibers, as a reference result, confirms this thesis.

3.2. Compressive Strength

The compressive strength test was carried out for concrete used for the production of block supports and concrete used to make a concrete slab. Below in Table 7 and Figure 15 are the results of tests of concrete with additives in the form of fibers and reference concrete.

Table 7. Results of the compressive strength test.

Concrete Type	Compressive Strength [MPa]	Standard Deviation [MPa]
Z1	59.80	2.12
Z2	71.67	3.26
Z3	72.87	1.14
Z4	76.98	1.52
Z5	68.50	2.14

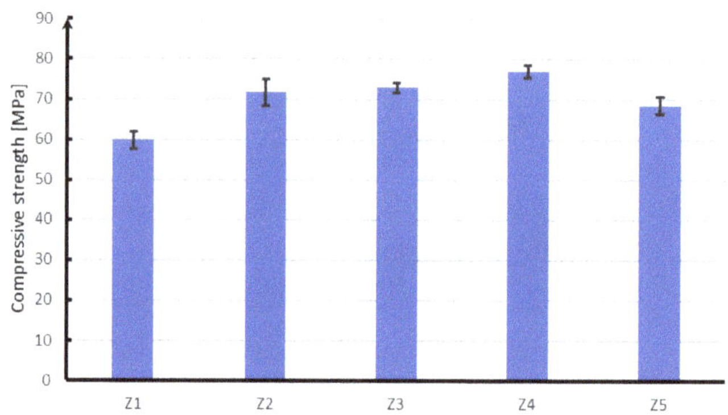

Figure 15. Results of the compressive strength test.

Based on the presented results, it can be concluded that the use of fibers slightly increased the compressive strength of concrete intended for block supports. The average strength of concrete without the addition of Z5 fibers is 68.5 MPa, while the strength of concrete with the addition of fibers is from 71.7 MPa for Z2 concrete, to 76.9 MPa for Z4 concrete, which is a strength increase of 4.7% to 12.3%. Although fibers have a greater impact on the increase in tensile strength than on compression, the conducted research shows that their use significantly increases the compressive strength and thus the class of concrete. In addition, it should be stated that the strength reserve is acceptable and allows us to assume that the concrete intended for the slab (Z1) and block supports (Z2–Z5) has been designed correctly.

3.3. Frost Resistance

The frost resistance test was carried out for concrete intended for the production of block supports and concrete intended for making a concrete slab. Below in Table 8 are the results of tests of concrete with additives in the form of fibers and reference concrete.

Based on the analysis of the obtained results, it should be concluded that all concrete samples meet the standard requirements. The permissible maximum loss in weight of samples after the frost resistance test should not exceed 5%, and the decrease in strength in relation to comparative samples not exposed to frost should not exceed 20%. All designed concretes are characterized by a loss of mass below 0.5% and a decrease in strength below

4%, which significantly exceeds the standard requirements. The research also shows that the use of fibers improved the frost resistance parameters of concrete intended for block supports in relation to the reference concrete Z5. In the case of steel fibers, the Z2 sample, the average strength decrease is reduced by 45.2% compared to the average strength decrease of concrete without fibers, and the average loss weight is reduced by 26.5%. For concrete with polymer fibers, Z3 concrete, the average strength decrease is reduced by 27.5%, and the average loss weight is reduced by 17.6%. When it comes to concrete with glass fibers, Z4 samples, the average strength decrease is reduced by 39% compared to the average strength decrease of concrete without fibers, and the average loss weight is reduced by 5.6%. Both due to the compressive strength and loss of mass after the impact of negative temperatures, concrete with the use of steel fibers looks the most advantageous. In terms of the decrease in compressive strength, Z4 concrete with glass fibers is not much worse, but in the case of a loss of mass, it is characterized by slightly more favorable results than concrete without fibers.

Table 8. Results of the frost resistance test.

Concrete Type	Average Weight Loss (%)	Compressive Strength of the Comparative Samples (MPa)	Compressive Strength of Samples after Frost Resistance Test (MPa)	Average Strength Decrease (%)
Z1	0.21	59.8	58.2	2.67
Z2	0.25	73.3	72.0	1.77
Z3	0.28	73.1	71.4	2.34
Z4	0.32	76.3	74.9	1.97
Z5	0.34	71.3	69.0	3.23

3.4. Modulus of Elasticity

As in the case of previous tests, the test of the modulus of elasticity was carried out for concrete intended for a concrete slab and concrete block supports with fibers and reference concrete without the addition of fibers. In the non-destructive method, based on the frequency of vibrations propagating in the sample, the modulus of elasticity was determined by the method of propagation of the wave in the sample. In the classical method, the modulus of elasticity was calculated on the basis of the recorded deformation of the sample during loading according to Formula (2). The results of the elastic modulus for individual samples determined by method one are summarized in Table 9 and by the classical method in Table 10.

$$E = \frac{\Delta\sigma}{\Delta\varepsilon} \quad (2)$$

where:

$\Delta\sigma$—compressive stress increment (MPa);

$\Delta\varepsilon$—sample deformation.

The results obtained by the resonant method are higher, but the difference between the results in relation to the classical method is about 2% to 4%. In both test methods used, the influence of the addition of fibers on the modulus of elasticity of cement concrete can be seen. The highest modulus of elasticity in both resonance and load tests is characterized by samples with glass fibers, whose modulus of elasticity is greater than concrete without fibers in the resonant method by 4.8% and in the classical method by 3.9%. Slightly worse are samples with polypropylene fibers, whose elastic modulus is higher by 4.2% in the resonance method and in the classical method by 3.8%. Steel fibers, whose elastic modulus is higher by 1.4% in the resonance method, and by 1.2% in the classical method, have the least positive effect.

Table 9. Results of the resonance test of the modulus of elasticity.

Concrete Type	Medium Modulus of Elasticity (MPa)	Standard Deviation (MPa)
Z1	40,467	550.8
Z2	45,433	115.5
Z3	46,667	288.7
Z4	46,933	208.2
Z5	44,800	300.0

Table 10. Results of the classical elastic modulus test.

Concrete Type	Medium Modulus of Elasticity (MPa)	Standard Deviation (MPa)
Z1	38,567	472.6
Z2	44,100	264.6
Z3	45,233	416.3
Z4	45,267	321.5
Z5	43,567	321.5

3.5. Tensile Strength of the Cross-Section of the Ballastless Railway Surface

In order to determine the influence of the strength parameters of block supports on the concrete slab, tensile strength tests at bending described in Section 2 of the laboratory model reflecting the actual track layout in the cross-section, with a scale factor of 0.3, were carried out. Z1 concrete was used as the material of the concrete slab and for block supports both concrete without Z5 fibers and with the addition of Z2 steel fibers, Z3 polypropylene and Z4 glass fibers. The results of the tensile strength at bending are shown in Table 11.

Table 11. Results of the tensile strength test at bending of the Z1 concrete beam depending on the reinforcing fibers used in the block supports.

Concrete Type	Average Tensile Strength (MPa)	Standard Deviation (MPa)
Z2	6.62	0.42
Z3	6.05	0.10
Z4	7.03	0.34
Z5	5.98	0.16

Based on the analysis of the presented results, it should be concluded that the fibers used in the concrete of block supports affect the tensile strength when bending the concrete slab. The highest tensile strength during bending has a plate with supports in which glass fibers are used, and the smallest with supports in which concrete without the addition of fibers is used. Glass fibers in concrete intended for block supports increased the tensile strength of the plate at bending by 17.5% compared to block supports in which no fibers were used. Steel fibers increased the tensile strength of the plate at bending by 10.7%, and polypropylene fibers by only 1.2%. It should also be noted that the degree to which individual fibers affect does not depend either on the compressive strength of the cement concrete or on the modulus of elasticity. The key in this case is how the material of block supports cooperates with the concrete slab in the transfer of loads.

3.6. *Tensile Strength of the Longitudinal Section of the Ballastless Railway Surface*

Another important study is to determine the influence of the material from which block supports were made on the tensile strength of the concrete slab in a longitudinal system. In this case, as before, the tests were performed on laboratory samples prepared on a scale of 0.3. The research was carried out for a concrete slab using block supports without fibers (Z5) and with the addition of fibers (Z2–Z4). The results of the study are presented in Table 12.

Table 12. Results of the tensile strength test at bending of the Z1 concrete beam depending on the reinforcing fibers used in the block supports.

Concrete Type	Average Tensil Strength (MPa)	Standard Deviation (MPa)
Z2	7.82	0.36
Z3	7.85	0.27
Z4	7.93	0.39
Z5	7.30	0.14

In the case of the longitudinal system, one can also notice increased tensile strength when bending the concrete slab after using reinforcing fibers in block supports. However, in the longitudinal arrangement, the difference in the results is not as large as in the case of the transverse arrangement. In the longitudinal system, block supports with glass fibers also had the best effect on increasing the tensile strength when bending the concrete slab and increased it by 8.6%. For the transverse system, it was 17.5%. Polypropylene and steel fibers increased the strength of the concrete slab by 7.5% and 7.1%, respectively. In the case of the longitudinal system, you can see a much more favorable effect of polypropylene fibers on the tensile strength when bending the concrete slab than in the case of the transverse system, where this benefit was only 1.2%. In the case of the longitudinal system, you can see a slightly more correlation of the impact of the increase in tensile strength when bending the concrete slab with the compressive strength of the concrete block supports. However, it is not possible to talk about a full correlation due to the value of standard deviations of the obtained results, which indicate that each type of fibers used equally affects the tensile strength of the concrete slab.

4. Discussion

On the basis of the conducted research, it can be concluded that individual strength results depend on the reinforcing fibers used. Samples with the addition of glass fibers are the most advantageous in the strength tests carried out, which are shown in Figures 16–19.

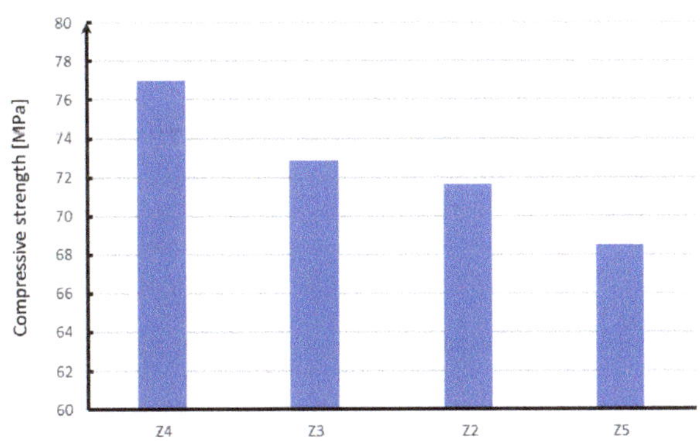

Figure 16. Results of the compressive strength test.

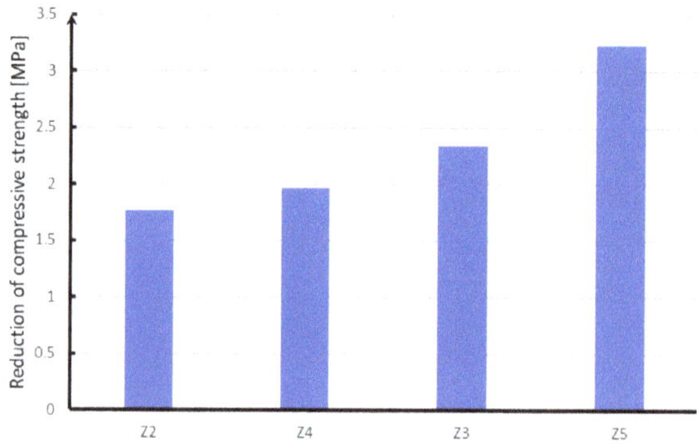

Figure 17. Results of the frost resistance test.

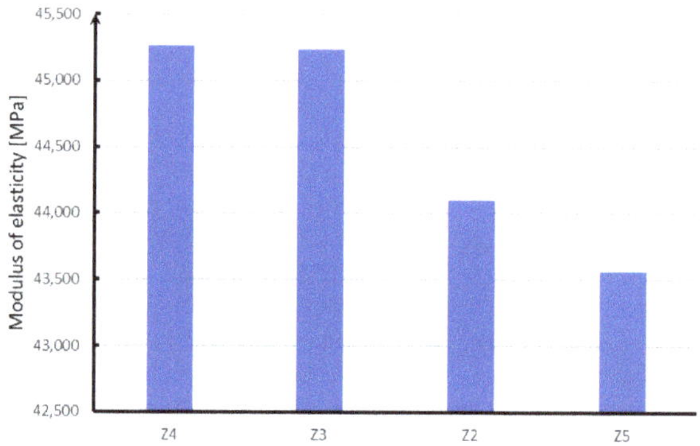

Figure 18. Results of the elastic modulus test.

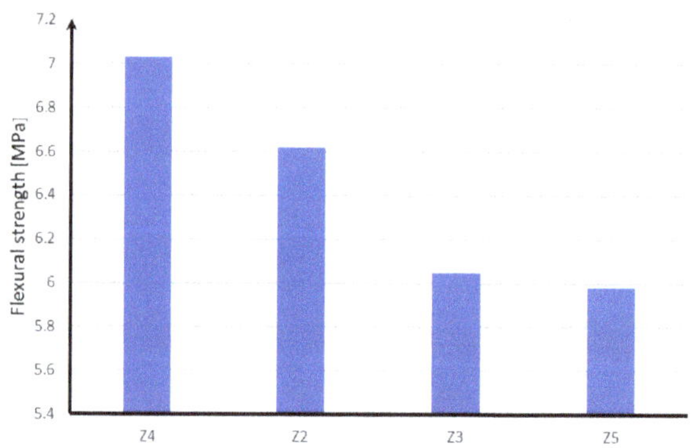

Figure 19. Stresses in the track plate depending on the fibers used in the block supports.

In the case of compression (Figure 16), the next in terms of strength are samples with the addition of polypropylene fibers, which in relation to glass fibers are characterized by a strength lower by about 4 MPa. On the other hand, steel fibers have a strength lower by about 5 MPa than glass fibers. The lowest compressive strength of about 68 MPa is characterized by concrete without reinforcing fibers.

In the case of frost resistance of cement concrete (Figure 17), samples with the addition of steel fibers, whose strength drop is 1.8 MPa, are the most advantageous. Glass fibers with the most favorable compressive strength are also characterized by the highest strength of samples subjected to frost, but their strength drop is greater than that of steel fibers and amounts to almost 2 MPa. Next are samples with the addition of polypropylene fibers, which are characterized by a decrease in strength equal to 2.3 MPa. The least favorable are samples without the addition of fibers, whose strength drop is over 3.2 MPa.

Analyzing the modulus of elasticity of concrete samples (Figure 18), it can be concluded that, as in the case of compressive strength, the highest module is characterized by samples with the addition of glass and polypropylene fibers. The difference between the obtained results for both fibers is within the limit of statistical error. Steel fibers are characterized by a modulus of elasticity lower by about 1000 MPa, while samples without the addition of fibers are characterized by a module lower by about 2000 MPa.

Taking into account the influence of reinforcing fibers on the work of the concrete slab supporting the block supports, it should be noted that the most advantageous variant is the use of glass fibers in block supports, thanks to which the concrete slab is characterized by the highest tensile strength in both longitudinal and transverse systems. In the case of steel fibers, the tensile strength of the concrete slab at bending is reduced by about 0.4 MPa. For polypropylene fibers, the strength is lower by about 1 MPa. The least favorable effect on the strength of the concrete slab is the use of block supports without the addition of fibers, which is shown in Figure 19.

The analyses carried out above show that the most advantageous in terms of strength and proper work of the surface is the use of glass fibers in block supports, which at the same time increase the strength of concrete intended for block supports and increase tensile strength when bending the concrete slab constituting support for block supports.

Having in mind the above discussion one may conclude what follows.

Tables 1–4 present the properties and compositions of concrete mixtures accepted for the production of samples without the addition of fibers (Z1 and Z5) and with the addition of fibers (Z2–Z4). All samples (Z1–Z5) were subjected to strength tests. Some of the tests concerned the samples themselves. The concrete mixes (Z1–Z5) were then used for elements of the railway surface, made on a laboratory scale. After that, additional tests were made for properly made aggregated railway track elements. The values of the tested physical quantities differed, and, moreover, their distribution was not the same in each individual test. The sequences of positions of individual material samples or elements made of them are different in different Tables 1–4 and different diagrams (Figures 16–19). Hence, the resulting material with various additives has different properties. This applies both to the compressive strength of the material samples themselves and to the bending strength of the fixed elements of the railway track, which are made of various materials with additives. The study of the influence of low temperatures on the change in the technical parameters of individual materials and elements made of them, as it was presented, is also very important. This factor in engineering practice must be taken into account, especially in climatic conditions, in which specific structures will be exposed to sub-zero temperatures or to cyclical temperature changes.

Finally, it should be stated that when changing the rail support in the transition railway sections, in order to eliminate the threshold effect, it is necessary to use a proper sequence of various reinforcing fibers in the set of railway concrete elements, which will properly affect the total rigidity of the track system in the transition zone.

5. Conclusions

This article is a continuation of the research described in [13], in which the phenomena occurring in the zones of changes in the technology of the railway surface were analyzed. In the transition zones from the ballast surface to the ballastless surface of the railway line, there are jumping changes in the strength parameters of the materials from which the elements of the surface are made. This adversely affects the dynamic interactions of the track–rail vehicle system. This phenomenon, called the threshold effect, was considered in [13], where reinforcements of materials on the ballast surface side were proposed, partially reducing the adverse effects of the threshold phenomenon. This paper presents how it is possible to influence the gradual change of strength parameters of materials on the ballastless side using concrete with various, gradually changing additives for the construction of pavement elements. As a consequence, the strength parameters in the transition zone, from ballast to ballastless surface at the entrance to the engineering facility and from ballastless to ballast surface when leaving the zone of the engineering facility, can be constructed in such a way that the change in the strength parameters of the materials used is not incremental, but rather changes continuously. It has been shown that this would be possible by using various concrete additives so that in a zone of appropriate length, the strength properties of the pavement elements change as continuously as possible.

The next step in the test cycle could be the use of an experimentally proven computational model [13], which would exploit the possibilities of affecting the ballast part of the surface in the transition zone [13] and the ballast-free part, discussed in this article, so as to propose design solutions in which the dynamic vehicle-track interactions will be limited as much as possible.

Author Contributions: Conceptualization, D.P. and W.I.; methodology, W.I., G.R. and T.R.; software, D.P. and T.L.; validation, W.I., G.R. and T.R.; formal analysis, W.I. and G.R.; investigation, D.P. and T.L.; writing—original draft preparation, W.I. and G.R.; writing—review and editing, W.I. and G.R.; visualization, G.R.; supervision, W.I. All authors have read and agreed to the published version of the manuscript.

Funding: This research was funded by the Faculty of Civil Engineering and Geodesy of the Military University of Technology: Warsaw, Poland—grant UGB number 794.

Institutional Review Board Statement: Not applicable.

Informed Consent Statement: Not applicable.

Data Availability Statement: The data presented in this study are available upon request from the corresponding author.

Acknowledgments: The authors would like to acknowledge the Authority of the Faculty of Civil Engineering and Geodesy of the Military University of Technology for providing administrative support during conducting our scientific work on this article.

Conflicts of Interest: The authors declare no conflict of interest.

References

1. Gołaszewski, J. *Domieszki do Betonu. Efekt Działania, Ocena i Badania Efektywności, Stosowanie*; Wydawnictwo Politechniki: Śląskiej, Gliwice, 2016.
2. Rudnicki, T. Natural and synthetic admixtures plasticize the tail and mechanisms of their interaction in the concrete mix. *Mag. Autostrady* **2004**, *4*, 22–25. [CrossRef]
3. Rudnicki, T. Functional Method of Designing Self-Compacting Concrete. *Materials* **2021**, *14*, 267. [CrossRef] [PubMed]
4. Glinicki, M.A. *Inżynieria Betonowych Nawierzchni Drogowych*; Wydawnictwo Naukowe PWN: Warszawa, Poland, 2019.
5. Pasławski, J.; Rudnicki, T. Agile/Flexible and Lean Management in Ready-Mix concrete delivery. *Arch. Civil Eng.* **2021**, *67*, 689–709. [CrossRef]
6. Rudnicki, T.; Jurczak, R. The impact of the addition of diabase dusts on the properties of cement pavement concrete. Arch. *Civil Eng.* **2022**, *68*, 395–411. [CrossRef]
7. Małek, M.; Jackowski, M.; Łasica, W.; Kadela, M. Influence of Polypropylene, Glass and Steel Fiber on the Thermal Properties of Concrete. *Materials* **2021**, *14*, 1888. [CrossRef] [PubMed]
8. Małek, M.; Łasica, Ł.; Kadela, M.; Kluczyński, J.; Dudek, D. Physical and Mechanical Properties of Polypropylene Fibre-Reinforced Cement–Glass Composite. *Materials* **2021**, *14*, 637. [CrossRef] [PubMed]
9. Foti, D. Innovative techniques for concrete reinforcement with polymers. *Constr. Build. Mater.* **2016**, *112*, 202–209. [CrossRef]
10. Jurczak, R.; Szmatuła, F.; Rudnicki, T.; Korentz, J. Effect of Ground Waste Glass Addition on the Strength and Durability of Low Strength Concrete Mixes. *Materials* **2021**, *14*, 190. [CrossRef] [PubMed]
11. Foti, D.; Lerna, M.; Sabbà, M.F.; Vacca, V. Mechanical characteristics and water absorption properties of blast-furnace slag concretes with fly ashes or microsilica additions. *Appl. Sci.* **2019**, *9*, 1279. [CrossRef]
12. Rudnicki, T.; Jurczak, R. Recycling of a Concrete Pavement after over 80 Years in Service. *Materials* **2020**, *13*, 2262. [CrossRef] [PubMed]

13. Idczak, W.; Lewandrowski, T.; Pokropski, D.; Rudnicki, T.; Trzmiel, J. Dynamic Impact of a Rail Vehicle on a Rail Infrastructure with Particular Focus on the Phenomenon of Threshold Effect. *Energies* **2022**, *15*, 2119. [CrossRef]
14. Vostroukhov, A.; Metrikine, A. Periodically supported beam on a visco-elastic layer as a model for dynamic analysis of a high-speed railway track. *Int. J. Solid Struct.* **2003**, *2003*, 5723–5752. [CrossRef]
15. Szafrański, M. *Wpływ Sposobu Odwzorowania Pojazdu Szynowego na Odpowiedz ´ Dynamiczną Przęsła Mostowego*; Infrastruktura Transportu Szynowego: Gdańsk, Poland, 2019.
16. Kaewunruen, S.; Lewandrowski, T.; Chamniprasart, K. Nonlinear modeling and analysis of moving train loads on interspersed railway tracks. In Proceedings of the 6th ECCOMAS Thematic Conference on Computational Methods in Structural Dynamics and Earthquake Engineering, Rhodes Island, Greece, 15–17 June 2017.
17. Kaewunruen, S.; Lewandrowski, T.; Chamniprasart, K. Dynamic responses of interspersed railway tracks to moving train loads. *Int. J. Struct. Stab. Dyn.* **2018**, *18*, 1850011. [CrossRef]
18. Goicolea, J.M.; Nguyen, K.; Galbadón, F.; Bermejo, M. Dynamic Analysis of High Speed Railway Traffic Loads on Ballast and Slab Tracks, Conference Paper January 2010. Available online: https://www.ctresources.info/ccp/paper.html?id=5725 (accessed on 23 June 2022).
19. Skaner Laserowy ScanCONTROL 2610-50-Micro-Epsilon. Available online: http://www.micro-epsilon.pl/produkt/10115/skaner-laserowy-scancontrol-2610-50/ (accessed on 23 June 2022).
20. Sołkowski, J.; Kudła, D. Analiza Niejednorodności Mechanicznych Nawierzchni i Podtorza w Obrębie Obiektu mostowego. Zeszyty Naukowo-Techniczne Stowarzyszenia Inżynierów i Techników Komunikacji Rzeczypospolitej Polskiej, Kraków. 2016. Available online: http://yadda.icm.edu.pl/baztech/element/bwmeta1.element.baztech-96c9ea1a-cdac-4031-85d5-243da1eb1405 (accessed on 23 June 2022).
21. Sołkowski, J. Zarys analizy efektu progowego przy łączeniu nawierzchni podsypkowych z innymi typami nawierzchni. *TTS Tech. Transp. Szyn.* **2009**, *15*, 59–65.
22. Sołkowski, J. *Efekt Progowy w Nawierzchniach Szynowych*; Politechnika Krakowska: Warszawa, Polska, 2013.
23. PN-EN 206.2014 Beton. Wymagania, Właściwości, Produkcja i Zgodność. Available online: https://sklep.pkn.pl/pn-en-206-2014-04p.html (accessed on 23 June 2022).
24. Specyfikacja Techniczna t.11.10.06 Bezpodsypkowa Konstrukcja Torowiska Tramwajowego w Postaci Prefabrykowanych Podkładów Dwublokowych Połączonych Konstrukcyjnie z Płytą Betonową. Available online: https://doczz.pl/doc/1998921/specyfikacja-techniczna-t.11.10.06-bezpodsypkowa-konstrukcja (accessed on 23 June 2022).
25. Specyfikacja Techniczna Wykonania i Odbioru Robót Budowlanych D-04.05.01 Podbudowa z Kruszywa Stabilizowanego Cementem. Available online: https://ozimek.pl/static/img/k01/ZP%20AK_2021/obek/SPECYFIKACJE%20TECHNICZNE%20WYKONANIA%20I%20ODBIORU%20ROB%C3%93T.pdf (accessed on 23 June 2022).
26. Specyfikacja Techniczna Wykonania i Odbioru Robót Budowlanych T-03.00.03. Budowa Nowej Nawierzchni Torów—Nawierzchnia Bezpodsypkowa. Gorzów Wielkopolski. Available online: https://www.gdansk.uw.gov.pl/attachments/article/1338/01_OST%20-%20WYMAGANIA%20OG%C3%93LNE.pdf (accessed on 23 June 2022).
27. Warunki Techniczne Wykonania i Odbioru Podkładów i Podrozjazdnic Strunobetonowych Id—101. Warszawa, Poland. 2010. Available online: https://www.plk-sa.pl/files/public/user_upload/pdf/Akty_prawne_i_przepisy/Instrukcje/Wydruk/Id/Id-101_WCAG.pdf (accessed on 23 June 2022).
28. PN-EN 14889-1:2007—Włókna do Betonu—Część 1: Włókna Stalowe—Definicje Wymagania i Zgodność. Available online: https://infostore.saiglobal.com/en-us/Standards/PN-EN-14889-1-2007-933809_SAIG_PKN_PKN_2200615/ (accessed on 23 June 2022).
29. Available online: www.siatpol-zbrojenia.pl (accessed on 23 June 2022).
30. PN-EN 14889—Włókna do Betonu. Available online: https://sklep.pkn.pl/pn-en-14889-1-2007p.html (accessed on 23 June 2022).
31. PN-EN 12350-2:2011 Badania Mieszanki Betonowej—Część 2: Badanie Konsystencji Metodą Opadu Stożka. Available online: https://sklep.pkn.pl/pn-en-12350-2-2011p.html (accessed on 23 June 2022).
32. PN-EN 12390-3:2019-08 Badania Betonu—Część 3: Wytrzymałość na Ściskanie Próbek do Badań. Available online: https://sklep.pkn.pl/pn-en-12390-3-2019-07p.html (accessed on 23 June 2022).
33. PN-88/B-06250 Beton Zwykły. Available online: https://sklep.pkn.pl/pn-b-06250-1988p.html (accessed on 23 June 2022).
34. PN-EN 12390-13-2014-02 Badania Betonu. Część 13: Wyznaczanie Siecznego Modułu Sprężystości Przy Ściskaniu. Available online: https://sklep.pkn.pl/pn-en-12390-13-2014-02e.html (accessed on 23 June 2022).

MDPI
St. Alban-Anlage 66
4052 Basel
Switzerland
Tel. +41 61 683 77 34
Fax +41 61 302 89 18
www.mdpi.com

Materials Editorial Office
E-mail: materials@mdpi.com
www.mdpi.com/journal/materials

www.ingramcontent.com/pod-product-compliance
Lightning Source LLC
Chambersburg PA
CBHW061548210926
51850CB00009B/2040
9783036580708